PEER INSTRUCTION
A USER'S MANUAL

ERIC MAZUR
Harvard University

PRENTICE HALL SERIES IN EDUCATIONAL INNOVATION

Prentice Hall
Upper Saddle River, New Jersey 07458

Library of Congress Cataloging-in-Publication Data

Mazur, Eric.
   Peer instruction: a user's manual/Eric Mazur.
     p. cm.
   Includes bibliographical references.
   ISBN 0-13-565441-6
   1. Physics—Study and teaching (Higher)   2. Peer-group tutoring of students. I. Title.
  QC30.M345 1997   96-20088
  530'.071--dc20
                                         CIP

Executive Editor: Alison Reeves
Editor-in-Chief: Paul F. Corey
Editorial Director: Tim Bozik
Development Editors: Irene Nunes and Carol Trueheart
Production Editor: Alison Aquino
Creative Director: Paula Maylahn
Cover Design: DeFranco Design, Inc.
Manufacturing Buyer: Trudy Pisciotti
Page Formatters: Eric Hulsizer and Jeff Henn
Copy Editor: Michael Schiaparelli

Photo credit, p.14, from *Thinking Together: Collaborative Learning in Science*, Derek Bok Center for Teaching and Learning, Harvard University, 1993.
Photo credit, p. 18, from David Meltzer and Kandiah Manivannan, Southeastern Louisiana University.
Photo credit, back cover, Jane Reed/Harvard University.

© 1997 by Eric Mazur
Published by Prentice Hall, Inc.
Upper Saddle River, New Jersey 07458

Printed in the United States of America.

20 19 18 17 16

ISBN 0-13-565441-6

Prentice Hall International (UK) Limited, *London*
Prentice Hall of Australia Pty. Limited, *Sydney*
Prentice Hall Canada, Inc., *Toronto*
Prentice Hall Hispanoamericana, S.A., *Mexico*
Prentice Hall of India Private Limited, *New Delhi*
Prentice Hall of Japan, Inc., *Tokyo*
Simon & Schuster Asia Pte. Limited, *Singapore*
Editora Prentice Hall do Brasil, Ltda., *Rio de Janeiro*

To my students, who taught me how to teach.

CONTENTS

FOREWORD by Charles W. Misner

The equation $F = ma$ is easy to memorize, hard to use, and even more difficult to understand. For most students, the ultimate value of an introductory physics course lies not in learning, say, this law of mechanics but in acquiring the skills a physicist uses in working with such laws. Important skills, transferable to other areas, include simplification, idealization, approximation, pictorial, graphical, and mathematical representations of phenomena and, more generally, mathematical/conceptual modeling. But the idea that physics is all equations and mathematics is such an established myth among students that many of them will refuse to think if they can find an equation to memorize as an alternative. Eric Mazur in this *Peer Instruction* guide shows teachers how to challenge students to think about the physics instead of juggling equations.

This is a very important step. If peer instruction methods are widely adopted, there could be a significant improvement in a large and important course. Designing lecture hours where students are active and interactive is an appealing idea. I've tried it after hearing such methods described by Mazur and by Thomas Moore. The results were encouraging but, I felt, limited by my ability (and time) to produce or find good discussion questions. I look forward to being able to use the high-quality *ConcepTests* to involve students and prevent their TV-viewing attention level from being switched on during the lecture hour.

A major advantage of Mazur's *User's Manual* is the supply of tested and adaptable materials to support the lecture designs he proposes. He supplies the full armory of necessary materials both in print and on CD-ROM. For example, the *ConcepTests* designed to provoke discussion and interaction during lectures will not be taken seriously by students unless conceptual questions appear on examinations. Consequently, this manual provides a large set of example questions that can be used or modified for examinations. Further, students will not be prepared for the interactive lecture (by reading) unless some pressure is brought to bear. Such pressure can be applied by using the large collection of reading quizzes made available for direct use or easy modification. The often deplored uniformity of introductory physics texts becomes an advantage here, since these materials can be used with all the main current texts.

In addition to providing the necessary materials, Mazur also gives a clear and detailed exposition of the methods he uses during the lecture hour in a large class setting. Since few schools can afford to teach introductory physics without the large lecture setting, these effective methods revitalize science teaching in ways that can be adopted immediately by the individual lecturer without the need for institutional reorganization. Mazur's approach will be equally accessible to the research scientist who also teaches and to the professor whose primary identification is with science education. I believe the publication of this manual is an important service to all physics teachers.

FOREWORD by Sheila Tobias

For ten years now, the physics community has been grappling with the problem of teaching introductory physics to undergraduates who come to class with "misconceptions" about the basic concepts. Building on research in cognitive science, leading physics educators have created new approaches, new demonstrations, interactive software, innovative pedagogies, and some new content to make the teaching of physics more effective and the study of physics more attractive to a wider group of undergraduate students.

Physics educators are now trying to disseminate some of these new undergraduate teaching approaches. Implementation, however, is a challenge. Even faculty with the wherewithal to attend national meetings and carry out small-scale experiments based on what they learned often encounter institutional constraints that limit the degree to which they can apply these new ideas to standard courses.

From my research and fieldwork,[1] I feel that while there is an important place for curriculum revisions, instructional software, and rich new pedagogy, instructors do not need an entirely new curriculum or complex (and expensive) pedagogical devices. Instead, instructors need an assist that enhances learning and, at the same time, provides a better teaching experience.

Eric Mazur's approach is, in essence, such an assist. His *Peer Instruction* manual offers an extraordinary, interactive approach to teaching physics which emphasizes understanding over and above what some of my experimental subjects call the *tyranny of technique*. His approach actively involves the students in the teaching process, making physics significantly more accessible to them. But how do you implement interactive teaching in large, heterogeneous classes?

---

[1]Sheila Tobias, *Revitalizing Undergraduate Science: Why Some Things Work and Most Don't*, Tucson, AZ: Research Corporation, (1992).

Here, for the first time in a user-friendly manual, is a step-by-step approach to teaching physics, that works. With this manual, physics educators have a guide to preparing interactive lectures for a one-year introductory physics course. (Chemistry and biology educators will also find much that is useable). The manual is organized by major topics, indexed and searchable using key words and concepts, and includes diagnostic tests, reading quizzes, and a full set of conceptual questions (*ConcepTests*) for class discussion.

Eric Mazur's *Peer Instruction* approach has been successfully field-tested in a variety of settings, most of them quite different from his home campus at Harvard University. At the University of Massachusetts, Lowell and at Appalachian State University, for instance, physics professors have found ways to employ both the *ConcepTests* and *Peer Instruction*. From my experience—and especially in this economic environment—instructors and their deans need something that provides a teaching improvement that can be implemented without substantial investment of either time or money, and that answers the question: "What can we do, today?"

The answer is *Peer Instruction*, in combination with *ConcepTests*, reading quizzes, and conceptual exam questions. As a model of useable material, this book breaks new ground. Faculty—and most particularly their students—will be grateful for these tools.

PREFACE

I love teaching. What attracted me to science was not only the excitement of doing science, the beauty of discovering new truths, but also the satisfaction of transferring this excitement and curiosity to others.

I have taught undergraduates at Harvard since joining the faculty in 1984. Initially I thought—as many other people do—that what is taught is learned, but over time I realized that nothing could be further from the truth. Analysis of my students' understanding of Newtonian mechanics made it clear: They were not all learning what I wanted them to learn. I could have blamed the students for this had I not always been bothered by the frustration that introductory science courses stir up in some students. What is it about science that can lead to such frustration? I decided to change my teaching style and discovered that I could do much better in helping my students learn physics. That is what this manual is about.

I have developed an interactive teaching style that helps students better understand introductory physics. The technique, named *Peer Instruction*, actively involves the students in the teaching process. It is simple and—as many others have demonstrated—it can easily be adapted to fit individual lecture styles. It makes physics not only more accessible for students but also easier to teach.

This manual contains a step-by-step guide on how to plan *Peer Instruction* lectures using your existing lecture materials. In addition, it includes a complete set of class-tested and ready-to-use material to implement the method in a one-year introductory physics course:

- Two diagnostic tests to evaluate your students' understanding of mechanics.
- Student questionnaire handouts to assess students' expectations for the course and to point out misconceptions.

- 44 Reading Quizzes, organized by subject and designed to be given at the beginning of each class to motivate the students to read assigned material before class.

- 243 *ConcepTests*, multiple-choice questions for use in lecture to engage the students and to assess their understanding.

- 109 Conceptual Examination Questions, organized by major topic and designed to reinforce the basic philosophy of the method of *Peer Instruction*.

The enclosed CD-ROM contains these same materials reformatted as necessary so that they can be easily reproduced as overhead transparencies or handouts. (See the Appendix on CD-ROM Instructions for more details.) The resources are a work in progress, and will continue to evolve. To complement the material in this book, a continually updated set of additional resources is available on the Project Galileo web site (http://galileo galileo.harvard.edu). In addition to resource material, the project Galileo web site provides a forum for the community of instructors who are implementing *Peer Instruction* in their courses. Your participation will be much appreciated by all users. I also welcome your comments, suggestions, or corrections for this manual. Please feel free to send me e-mail at "mazur@physics.harvard.edu".

Many have contributed to this effort. The idea of using questions during the lecture was first suggested to me by Dudley Herschbach in the Chemistry Department at Harvard University. Debra Alpert, who joined me as a postdoctoral associate in 1991, has assisted me with much of the research here and played an active role in the development of the resource material. I am grateful to Anne Hoover who distributed hundreds of copies of an early version of this manual, allowing many people to introduce the method at their own institutions. I thank them all. I would also like to thank my colleagues Michael Aziz, William Paul, and Robert M. Westervelt at Harvard for their willingness to experiment along with me and for their contributions to the resource material. All of us owe much to the students in Physics 11 at Harvard, who were an integral part of the early experiments and who taught us how to teach them. I would also like to thank Albert Altman for his unfailing enthusiasm and the energy with which he implemented the method at the University of Massachusetts at Lowell and Charles Misner for the excellent suggestion to include resource material with the manual. Special thanks go to David Hestenes, Ibrahim Halloun, Eugene Mosca, Richard Hake, the late Malcolm Wells and Gregg Swackhamer for developing the *Force Concept Inventory* and the *Mechanics Baseline Test* as well as for their permission to include these in the book.

I am enormously grateful to the following reviewers of the manuscript for *Peer Instruction: A User's Manual* and their many insightful and pragmatic comments: Albert Altman, University of Massachusetts, Lowell; Arnold Arons, University of Washington; Bruce B. Birkett II, University of California, Berkeley; Paul Draper, University of Texas at Arlington; Robert J. Endorf, University of

Cincinnati; Thomas Furtak, Colorado School of Mines; Ian R. Gatland, Georgia Institute of Technology; J. David Gavenda, University of Texas at Austin; Kenneth A. Hardy, Florida International University; Greg Hassold, GMI Engineering and Management Institute; Peter Heller, Brandeis University; Laurent Hodges, Iowa State University; Mark W. Holtz, Texas Tech University; Zafir A. Ismail, Daemen College; Arthur Z. Kovacs, Rochester Institute of Technology; Dale D. Long, Virginia Polytechnic Institute; John D. McCullen, University of Arizona; James McGuire, Tulane University; Charles W. Misner, University of Maryland, College Park; George W. Parker, North Carolina State University; Claude Penchina, University of Massachusetts, Amherst; Joseph Priest, Miami University; Joel R. Primack, University of California, Santa Cruz; Lawrence B. Rees, Brigham Young University; Carl A. Rotter, West Virginia University; Leonard Scarfone, University of Vermont; Leo J. Schowalter, Rensselaer Polytechnic Institute; H. L. Scott, Oklahoma State University; Shahid A. Shaheen, Florida State University; Roger L. Stockbauer, Louisiana State University; William G. Sturrus, Youngstown State University; Robert S. Weidman, Michigan Technological University.

Finally I would like to thank Tim Bozik at Prentice Hall for encouraging me to publish this manual and Irene Nunes, who edited this manuscript with meticulous attention to detail and who contributed many valuable comments. I am also grateful to Alison Reeves, Alison Aquino, Carol Trueheart, Ray Mullaney, Eric Hulsizer, and Jeff Henn who all worked hard to turn the manuscript into a book.

CONCORD, MA

This work was partially supported by the Pew Charitable Trusts and by the National Science Foundation under contracts USE-9156037 and DUE-9254027.

This project was supported, in part,
by the
National Science Foundation
Opinions expressed are those of the authors
and not necessarily those of the Foundation

PEW
SCIENCE PROGRAM
IN UNDERGRADUATE EDUCATION

PART ONE

OVERVIEW

1

INTRODUCTION

The introductory physics course often is one of the biggest hurdles in the academic career of a student. For a sizable number of students, the course leaves a permanent sense of frustration. I have only to tell people I am a physicist to hear grumblings about high school or college physics. This general sense of frustration with introductory physics is widespread among non-physics majors required to take physics courses. Even physics majors are frequently dissatisfied with their introductory courses, and a large fraction of students initially interested in physics end up majoring in a different field. Why does this happen, and can we do something about it? Or should we just ignore this phenomenon and concentrate on teaching the successful student who is going on to a career in science?

AN EYE OPENER

Frustration with introductory physics courses has been commented on since the days of Maxwell and has recently been widely publicized by Sheila Tobias, who asked a number of graduate students in the humanities and social sciences to audit introductory science courses and describe their impressions.[1] The result of this survey is a book that paints a bleak picture of introductory science education. One may be tempted to brush off complaints by non-physics majors as coming from students who are *a priori* not interested in physics. Most of these students, however, are not complaining about other required courses outside their major field. In science education, in Tobias' words, the next generation of science workers is expected to rise like cream to the top, and the system is unapologetically competitive, selective, and intimidating, designed to winnow out all but the top tier.

[1] Sheila Tobias, *They're Not Dumb, They're Different: Stalking the Second Tier,* Tucson, AZ: Research Corporation, (1990).

The way physics is taught in the 1990s is not much different from the way it was taught—to a much smaller and more specialized audience—in the 1890s, and yet the audience has vastly changed. Physics has become a building block for many other fields, and enrollment in physics courses has grown enormously, with the majority of students not majoring in physics. This shift in constituency has caused a significant change in student attitude toward the subject and made the teaching of introductory physics a considerable challenge. Although conventional methods of physics instruction have produced many successful scientists and engineers, far too many students are unmotivated by the conventional approach. What, then, is wrong with it?

I have been teaching an introductory physics course for engineering and science majors at Harvard University since 1984. Until 1990, I taught a conventional course consisting of lectures enlivened by classroom demonstrations. I was generally satisfied with my teaching—my students did well on what I considered difficult problems, and the evaluations I received from them were very positive. As far as I knew, there were not many problems in *my* class.

In 1990, however, I came across a series of articles by Halloun and Hestenes[2] that really opened my eyes. As is well known, students enter their first physics course possessing strong beliefs and intuitions about common physical phenomena. These notions are derived from personal experience and color students' interpretations of material presented in the introductory course. Halloun and Hestenes show that instruction does little to change these "commonsense" beliefs.

For example, after a couple of months of physics instruction, all students can recite Newton's third law and most of them can apply it in numerical problems. A little probing, however, quickly shows that many students do not understand the law. Halloun and Hestenes provide many examples in which students are asked to compare the forces exerted by different objects on one another. When asked, for instance, to compare the forces in a collision between a heavy truck and a light car, many students firmly believe the heavy truck exerts a larger force. When reading this, my first reaction was "Not *my* students…!" Intrigued, I decided to test my own students' conceptual understanding, as well as that of the physics majors at Harvard.

The first warning came when I gave the Halloun and Hestenes test to my class and a student asked, "Professor Mazur, how should I answer these questions? According to what you taught us, or by the way I *think* about these things?" Despite this warning, the results of the test came as a shock: The students fared hardly better on the Halloun and Hestenes test than on their midterm examination. Yet, the Halloun and Hestenes test is *simple*, whereas the material covered by the examination (rotational dynamics, moments of inertia) is of far greater difficulty or so I thought.

---

[2] Ibrahim Abou Halloun and David Hestenes, *Am. J. Phys*, 53, (1985), 1043; *ibid*. 53, (1985), 1056 ; *ibid.* 55, (1987), 455; David Hestenes, *Am. J. Phys*, 55, (1987), 440.

MEMORIZATION VERSUS UNDERSTANDING

To understand these seemingly contradictory observations, I decided to pair, on subsequent examinations, simple qualitative questions with more difficult quantitative problems on the same physical concept. An example of a set of such questions on dc circuits is shown in Figure 1.1. These questions were given as the first and last problem on a midterm examination in the spring of 1991 in a conventionally taught class (the other three problems on the examination, which were placed between these two, dealt with different subjects and are omitted here).

---

1. A series circuit consists of three identical light bulbs connected to a battery as shown here. When the switch S is closed, do the following increase, decrease, or stay the same?

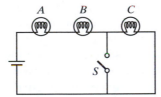

(a) The intensities of bulbs A and B
(b) The intensity of bulb C
(c) The current drawn from the battery
(d) The voltage drop across each bulb
(e) The power dissipated in the circuit

5. For the circuit shown, calculate (a) the current in the 2-Ω resistor and (b) the potential difference between points P and Q.

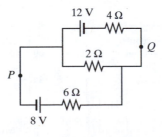

Figure 1.1 Conceptual (top) and conventional question (bottom) on the subject of dc circuits. These questions were given on a written examination in 1991.

Notice that question 1 is purely conceptual and requires only a knowledge of the fundamentals of simple circuits. Question 5 probes the students' ability to deal with the same concepts, now presented in the conventional numerical format. It requires setting up and solving two equations using Kirchhoff's laws. Most physicists would consider question 1 easy and question 5 harder. As the result in Figure 1.2 indicates, however, students in a conventionally taught class would disagree.

Analysis of the responses reveals the reason for the large peak at 2 for the conceptual question: Over 40% of the students believed that closing the switch doesn't change the current through the battery but that the current splits into two at the top junction and rejoins at the bottom! In spite of this serious misconception, many still managed to correctly solve the mathematical problem.

Figure 1.3 shows the lack of correlation between scores on the conceptual and conventional problems of Figure 1.1. Although 52% of the scores lie on the broad diagonal band, indicating that these students achieved roughly equal scores on both questions (±3 points), 39% of the students did substantially worse on the conceptual question. (Note that a number of students managed to score zero on the conceptual question and 10 on the conventional one!) Conversely, far fewer students (9%) did worse on the conventional question. This trend was confirmed on many similar pairs of problems during the remainder of the semester: Students tend to perform significantly better when solving standard textbook problems than when solving conceptual problems covering the same subject.

This simple example exposes a number of difficulties in science education. First, it is possible for students to do well on conventional problems by memorizing algorithms without understanding the underlying physics. Second, as a result of this, it is possible for a teacher, even an experienced one, to be completely misled into thinking that students have been taught effectively. Students are subject to the same misconception: They believe they have mastered the material and then are severely frustrated when they discover that their plug-and-chug recipe doesn't work in a different problem.

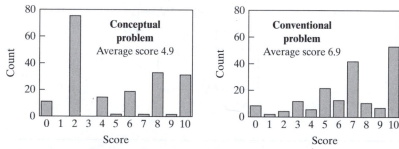

Figure 1.2 Test scores for the problems shown in Figure 1.1. For the conceptual problem, each part was worth a maximum of 2 points.

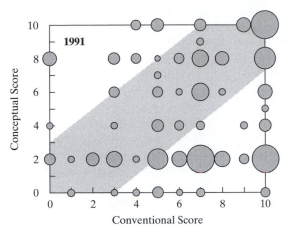

Figure 1.3 Correlation between conceptual and conventional problem scores from Figure 1.2. The radius of each datapoint is a measure of the number of students represented by that point.

Clearly, many students in my class were concentrating on learning "recipes," or "problem-solving strategies" as they are called in textbooks, without considering the underlying concepts. Plug and chug! Many pieces of the puzzle suddenly fell into place:

- The continuing requests by students that I do more and more problems and less and less lecturing—isn't this what one would expect if students are tested and graded on their problem-solving skills?

- The inexplicable blunders I had seen from apparently bright students—problem-solving strategies work on some but surely not on all problems.

- Students' frustration with physics—how boring physics must be when it is reduced to a set of mechanical recipes that do not even work all the time!

2

PEER INSTRUCTION

One problem with conventional teaching lies in the presentation of the material. Frequently, it comes straight out of textbooks and/or lecture notes, giving students little incentive to attend class. That the traditional presentation is nearly always delivered as a monologue in front of a passive audience compounds the problem. Only exceptional lecturers are capable of holding students' attention for an entire lecture period. It is even more difficult to provide adequate opportunity for students to critically think through the arguments being developed. Consequently, lectures simply reinforce students' feelings that the most important step in mastering the material is solving problems. The result is a rapidly escalating loop in which the students request more and more example problems (so they can learn better how to solve them), which in turn further reinforces their feeling that the key to success is problem-solving.

WHY LECTURE?

The first time I taught introductory physics, I spent a lot of time preparing lecture notes, which I would then distribute to my students at the end of each lecture. The notes became popular because they were concise and provided a good overview of the much more detailed information in the textbook.

Halfway through the semester, a couple of students asked me to distribute the notes in advance so they would not have to copy down so much and could pay more attention to my lecture. I gladly obliged, and the next time I was teaching the same course, I decided to distribute the collected notes all at once at the beginning of the semester. The unexpected result, however, was that a number of students complained on their end-of-semester questionnaires that I was lecturing straight out of my lecture notes!

Ah, the ungratefulness! I was at first disturbed by this lack of appreciation but have since changed my position. The students had a point: I was indeed lecturing from my lecture notes. And research showed that my students were deriving little additional benefit from hearing me lecture if they had read my notes beforehand. Had I lectured not on physics but, say, on Shakespeare, I would certainly not spend the lectures reading plays to the students. Instead, I would ask the students to read the plays before coming to the lecture and I would use the lecture periods to discuss the plays and deepen the students' understanding of and appreciation for Shakespeare.

In the years following the eye-opening experience described in Chapter 1, I explored new approaches to teaching introductory physics. In particular, I was looking for ways to focus attention on the underlying concepts without sacrificing the students' ability to solve problems. The result is *Peer Instruction*, an effective method that teaches the conceptual underpinnings in introductory physics and leads to better student performance on conventional problems. Interestingly, I have found this new approach also makes teaching easier and more rewarding.

The improvements I have achieved with *Peer Instruction* require the textbook and the lectures to play roles different from those they play in a conventional course. Preclass reading assignments from the book first introduce the material. Next, lectures elaborate on the reading, address potential difficulties, deepen understanding, build confidence, and add additional examples. Finally, the book serves as a reference and a study guide.

THE CONCEPTEST

The basic goals of *Peer Instruction* are to exploit student interaction during lectures and focus students' attention on underlying concepts. Instead of presenting the level of detail covered in the textbook or lecture notes, lectures consist of a number of short presentations on key points, each followed by a *ConcepTest*—short conceptual questions on the subject being discussed. The students are first given time to formulate answers and then asked to discuss their answers with each other. This process (*a*) forces the students to think through the arguments being developed, and (*b*) provides them (as well as the teacher) with a way to assess their understanding of the concept.

Each *ConcepTest* has the following general format:
1. Question posed   1 minute
2. Students given time to think   1 minute
3. Students record individual answers (optional)
4. Students convince their neighbors (peer instruction)   1–2 minutes
5. Students record revised answers (optional)
6. Feedback to teacher: Tally of answers
7. Explanation of correct answer   2+ minutes

If most students choose the correct answer to the *Concep Test*, the lecture proceeds to the next topic. If the percentage of correct answers is too low (say less than 90%), I slow down, lecture in more detail on the same subject, and reassess with another *Concep Test*. This repeat-when-necessary approach prevents a gulf from developing between the teacher's expectations and the students' understanding—a gulf that, once formed, only increases with time until the entire class is lost.

Let's consider a specific example: Archimedes' principle. I first lecture for 7–10 minutes on the subject—emphasizing the concepts and the ideas behind the proof while avoiding equations and derivations. This short lecture period could include a demonstration (the Cartesian diver, for instance). Then, before proceeding to the next topic (Pascal's principle, perhaps), I use the overhead projector to show the question presented in Figure 2.1.

I read the question to the students, making sure there are no misunderstandings about it. Next, I tell them they have one minute to select an answer—more time allows them to fall back onto equations rather than think. Since I want each student to answer individually, I do not allow them to talk to one another; I make sure it is dead-silent in the classroom. After about a minute, I ask the students first to record their answer and then to try to convince a neighbor

BUOYANCY

Imagine holding two identical bricks under water. Brick *A* is just beneath the surface of the water, while brick *B* is at a greater depth. The force needed to hold brick *B* in place is

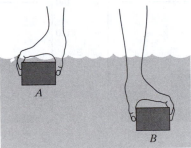

1. larger than
2. the same as
3. smaller than

the force required to hold brick *A* in place.

Figure 2.1 *Concep Test* question on Archimedes' principle. For an incompressible fluid such as water, the second choice is correct.

of the rightness of that answer. I always participate with a few groups of students in the animated discussions that follow. Doing so allows me to assess the mistakes being made and to hear how students who have the right answer explain their reasoning. After giving the students a minute or so to discuss the question, I ask them to record a revised answer. Then I return to the overhead projector and ask for a show of hands to check the distribution of answers. The results for the question in Figure 2.1 are shown in Figure 2.2. Of course, I did not have the detailed results shown in Figure 2.2 available in class, but a show of hands after the convince-your-neighbors discussion revealed an overwhelming majority of correct answers. I therefore spent only a few minutes explaining the correct answer before going on to the next topic.

The convince-your-neighbors discussions systematically increase both the percentage of correct answers and the confidence of the students. The improvement is usually largest when the initial percentage of correct answers is around 50%. If this percentage is much higher, there is little room for improvement; if it is much lower, there are too few students in the audience to convince others of the correct answer. This finding is illustrated in Figure 2.3, which shows the improvement in correct responses and confidence for all questions given during a semester. Notice that all points lie above a line of slope 1 (for points below that line, the percentage of correct responses after discussion is decreased). I consider an initial percentage of correct responses in the 40 to 80% range optimal and, in subsequent semesters, usually modify or eliminate questions that fall outside this range.

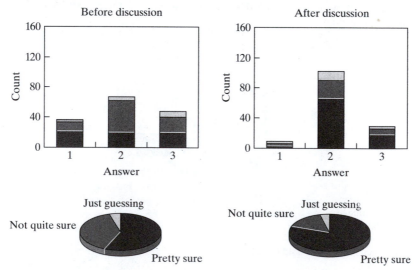

Figure 2.2 Data analysis of responses to the buoyancy question of Figure 2.1. The correct answer is 2. The pie charts show the overall distribution in confidence levels, and the shading in the bars corresponds to the shadings defined in the pie charts.

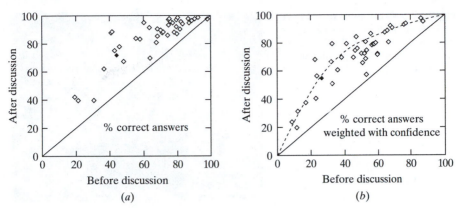

Figure 2.3 (*a*) Percentage of correct answers after discussion versus percentage before discussion and (*b*) the same information weighted with the students' confidence. Each datapoint corresponds to a single *ConcepTest* question. The filled datapoint is for the buoyancy question in Figure 2.1.

Figure 2.4 shows how students revised their answers in the discussion of the buoyancy question posed in Figure 2.1. In fact, 29% correctly revised their initially incorrect answer, while only 3% changed from correct to incorrect. Figure 2.3 demonstrates that there is always an increase and never a decrease in the percentage of correct answers. The reason is that it is much easier to change the mind of someone who is wrong than it is to change the mind of someone who has selected the right answer for the right reasons. The observed improvement in confidence is also no surprise. Students who are initially right but not very confident become more confident when it appears that neighbors have chosen the same answer or when their confidence is reinforced by reasoning that leads to the right answer.

At times, it seems that students are able to explain concepts to one another more effectively than are their teachers (see Figure 2.5). A likely explanation is that students who understand the concept when the question is posed have only recently mastered the idea and are still aware of the difficulties involved in grasping that concept. Consequently, they know precisely what to emphasize in

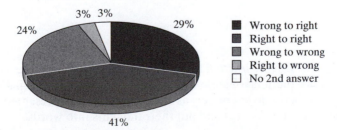

Figure 2.4 How answers were revised after convince-your-neighbors discussion for the buoyancy question in Figure 2.1.

Figure 2.5 *Peer Instruction* at work: students teaching students in a large introductory physics class. Nothing clarifies ideas better than explaining them to others.

their explanation. Similarly, many seasoned lecturers know that their first presentation of a new course is often their best, marked by a clarity and freshness often lacking in later, more polished versions. The underlying reason is the same: As time passes and a lecturer is continuously exposed to the material, the conceptual difficulties seem to disappear and therefore become harder to address.

In this new lecturing format, the *ConcepTests* take about one third of each lecture period, leaving less time for straight lecturing. One therefore has two choices: (*a*) discuss in lecture only part of the material to be covered over the span of the semester or (*b*) reduce the number of topics covered during the semester. While in some cases (*b*) may be the preferable choice, I have opted for (*a*): I do not cover in class all the material covered in the text and in the lecture notes that I pass out at the beginning of the term. I start by eliminating from my lectures worked examples and nearly all derivations. To make up for the omission of these mechanical details, I require the students to read the textbook and my lecture notes before coming to class. While this may sound surprising for a science course, students are accustomed to reading assignments in many other courses. In this way, students are exposed, over the length of the course, to the same amount of material taught in the conventional course.

Before getting into the specifics of *Peer Instruction*, let me summarize some of the striking results I have obtained—results supported by findings from other institutions where *Peer Instruction* has been implemented.[1]

RESULTS

The advantages of *Peer Instruction* are numerous. The convince-your-neighbors discussions break the unavoidable monotony of passive lecturing, and, more important, the students do not merely assimilate the material presented to them; they must think for themselves and put their thoughts into words.

[1] See Sheila Tobias, *Revitalizing Undergraduate Science Education: Why Some Things Work and Most Don't*, Tucson, AZ: Research Corporation, (1992).

To assess my students' learning, I have used two diagnostic tests since 1990: the *Force Concept Inventory* and the *Mechanics Baseline Test* (see Chapters 7 and 8).[2,3] The results of this assessment are shown in Figure 2.6 and Figure 2.7 and in Table 2.1. Figure 2.6 shows the dramatic gain in student performance obtained on the *Force Concept Inventory* when I first implemented *Peer Instruction* in 1991. As Table 2.1 shows, this gain was reproduced in subsequent years. Notice also how, in the posttest in Figure 2.6, the scores strongly shift toward full marks (29 out of 29) and that only 4% of the students remain below the cutoff identified by Hestenes as the threshold for the understanding of Newtonian mechanics. With the conventional approach (Figure 2.7), the gain was only half as large, in agreement with what has been found at other institutions for conventionally taught courses.

DO PROBLEM-SOLVING SKILLS SUFFER?

While improvement in conceptual understanding is undeniable, one might question how effective the new approach is in teaching the problem-solving skills required on conventional examinations. After all, restructuring of the lecture and its emphasis on conceptual material are achieved at the expense of lecture time devoted to problem-solving. Development of problem-solving skills is left to homework assignments and discussion sections.

A partial answer to this question can be obtained by looking at the scores for the *Mechanics Baseline* test, which involves some quantitative problem-solving. Table 2.1 shows that the average score on this test increased from 67% to 72% the year *Peer Instruction* was first used and rose to 73% and 76% in subsequent years.

[2] D. Hestenes, M. Wells, and G. Swackhamer, *Phys. Teach.* 30, (1992), 141.
[3] D. Hestenes, M. Wells, and G. Swackhamer, *Phys. Teach.* 30, (1992), 159.

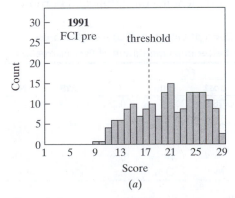

(a)

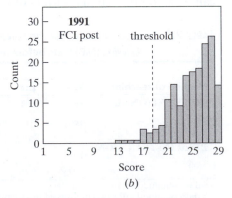

(b)

Figure 2.6 *Force Concept Inventory* scores obtained in 1991 (*a*) on the first day of class and (*b*) after two months of instruction with the *Peer Instruction* method. The maximum score on the test is 29. The means of the distributions are 19.8 (out of 29) for (*a*) and 24.6 for (*b*).

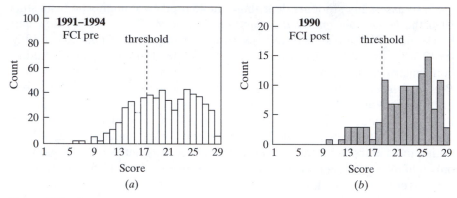

Figure 2.7   *Force Concept Inventory* scores obtained (*b*) in 1990 after two months of conventional instruction. For comparison, data obtained (*a*) on the first day of class in 1991, 1992, and 1994. The means of the distributions are 19.8 out of 29 points for (*a*) and 22.3 for (*b*).

For an unambiguous comparison of problem-solving skills on conventional examinations with and without *Peer Instruction*, I gave my 1985 final examination in 1991. Figure 2.8 shows the distributions of final examination scores for the two years. Given the students' improvement in conceptual understanding, I would have been satisfied if the distributions were the same. Instead, there is a marked improvement in the mean, as well as a higher cut-off in the low-end tail. Apparently, and perhaps not surprisingly, a better understanding of the underlying concepts leads to improved performance on conventional problems.

FEEDBACK

One of the great advantages of *Peer Instruction* is that the *ConcepTest* answers provide immediate feedback on student understanding. Tallying the answers can be accomplished in a variety of ways, depending on setting and purpose:

TABLE 2.1   Average scores for the *Force Concept Inventory* (FCI) and *Mechanics Baseline* (MB) tests before and after implementation of *Peer Instruction*

Method of teaching	Year	FCI pre[a]	post[b]	gain	G[c]	MB	N[d]
CONVENTIONAL	1990	(70%)[e]	78%	8%	0.25	67%	121
PEER INSTRUCTION	1991	71%	85%	14%	0.49	72%	177
	1993[f]	70%	86%	16%	0.55	73%	158
	1994	70%	88%	18%	0.59	76%	216
	1995[g]	67%	88%	21%	0.64	76%	181

[a] data obtained on first day of class
[b] data obtained after two months of instruction
[c] fraction of maximum possible gain realized
[d] number of data points
[e] no FCI pretest in 1990; 1991–1995 average shown
[f] no tests administered in 1992
[g] data for 1995 reflect use of manuscript for forthcoming text

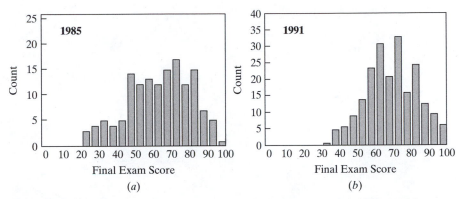

Figure 2.8 Final examination scores on identical final examination given (*a*) in 1985 (conventional course) and (*b*) in 1991 (*Peer Instruction*). The means of the distributions are 62.7 out of 100 for (*a*) and 69.4 for (*b*).

1. *Show of hands.* A show of hands after students have answered a question for the second time is the simplest method. It gives a feel for the level of the class' understanding and allows the teacher to pace the lecture accordingly. The main drawback is a loss of accuracy, in part because some students may hesitate to raise their hands and in part because of the difficulty in estimating the distribution. A nice solution is the use of so-called "flashcards"—each student has a set of six or more cards labeled A–F to signal the answer to a question (see Figure 2.9).[4] Other shortcomings are the lack of a permanent record (unless one keeps data in class) and the lack of any data collected before the convince-your-neighbors discussion (a show of hands before the discussion influences the outcome).

2. *Scanning forms.* Because I am interested in quantifying the effectiveness of convince-your-neighbors discussion in both the short and the long term, I have made extensive use of forms that I scan after class. On these forms, the students mark their answers and their confidence level, both before and after discussion. This method yields an enormous body of data on attendance, understanding, improvement, and the short-term effectiveness of the *Peer Instruction* periods. The drawbacks are that it requires some work after each lecture and that there is a delay in feedback, the data being available only after the forms are scanned. For this reason, when using scanning forms, I always also ask for a show of hands after the question is answered for the second time.

3. *Handheld computers.* Since 1993, I have used an interactive computer response system called *Classtalk*, produced by *Better Education, Inc.*[5] The system allows students to enter their answers to the *ConcepTests*, as well as their confidence levels, on a variety of handheld devices, ranging from graphing calculators to palmtop or laptop computers, that they share in small groups of three or

---

[4] David E. Meltzer and Kandiah Manivannan, *Phys. Teach.* 34, (1996), 72.

[5] Better Education, Inc., 4824 George Washington Memorial Highway, Suite 103, Grafton, VA 23692. Phone: (804) 898-4846. FAX: (804) 898-1897. Electronic mail: info@bedu.com

Figure 2.9   Students in an introductory physics class at Southeastern Louisiana University using flashcards—a simple, easy-to-implement, and cost-effective response system.

four. Their responses are relayed to the teacher on a computer screen and can be projected so the students see it, too. The main advantage of the system is that analysis of the results is available immediately. In addition, student information (such as name and seat location) is available to the instructor, making large classes more personal; the system can also handle numerical and non-multiple-choice questions; and sharing these hand-held computers enhances student interaction. Potential drawbacks are that the system requires a certain amount of capital investment and that it adds complexity to the lecture.

It is important to note that the success of *Peer Instruction* is *independent* of feedback method and therefore independent of financial or technological resources.

CONCLUSION

Using the lecture format described here, it is possible, with relatively little effort and no capital investment, to greatly improve student performance in introductory science courses—to double their gain in understanding as measured by the *Force Concept Inventory* test and improve their performance on conventional examinations. Despite the reduced time devoted to problem-solving, the results convincingly show that conceptual understanding enhances student performance on conventional examinations. Similar benefits have been obtained in a variety of academic settings with vastly different student bodies.[6] Finally, student surveys show that student satisfaction—an important indicator of student success—increases as well.

[6] Sheila Tobias, *Revitalizing Undergraduate Science Education: Why Some Things Work and Most Don't*, Tucson, AZ: Research Corporation, (1992). Also: R.R. Hake, *AAPT Announcer* 24 (2), (1994) 55; "Interactive-Engagement vs. Traditional Methods: A Six-Thousand-Student Survey of Mechanics Test Data for Introductory Physics Courses," preprint, June, 1995.

3

MOTIVATING THE STUDENTS

Students are not likely to accept a change in lecture format with open arms. They are used to traditional lectures and will doubt the new format will help them achieve more (*i.e.*, obtain a higher grade in the course). Since full student collaboration is essential to the success of the *Peer Instruction* method, it is important to motivate students early on. Here I outline what I have done in my class to win the students over.

SETTING THE TONE

I start on the first day of class by announcing to the students that I will not lecture straight out of my notes or out of the textbook. I argue that it would be a waste of their time to have me simply repeat what is printed in the textbook or the notes. To do so implies they are unable to read, and they ought to be offended when an instructor treats them that way!

I often quote the anecdote about my lecture notes on page 9 and explain how lecture time will be used and why they will benefit from it. I explain how little one tends to learn in a passive lecture, emphasizing that it is not possible for an instructor just to pour knowledge into their minds, that no matter how good the instructor, *they* still have to do the work. I challenge them to become critical thinkers, explaining the difference between plugging numbers into equations and being able to analyze an unfamiliar situation.

To drive this last point home, I announce to the students that they will be able to use a formula sheet on the examinations.[1] The formula sheet discourages memorization of equations and allows the students to focus on the meaning of the equations.

At the end of the first lecture, I give out the questionnaire reproduced in Figure 3.1. In spite of my introductory explanation of the way I will teach, the responses to this questionnaire show that a sizable fraction of students still expect to be lectured in the traditional way.

It is very important to make sure the students' expectations conform better to what will actually happen in class. In the second lecture, I therefore hand out a sheet with the results of the questionnaire (Chapter 9) and use up to fifteen minutes of lecture time to discuss it with the students. These fifteen minutes more than pay for themselves in the long run because better motivated students are more likely to become active participants in the learning process.

After about four weeks of lecturing, I issue the questionnaire shown in Figure 3.2. While I do not issue a detailed analysis of this questionnaire, I look for two things when reading the responses. First, is there something I can do to

---

[1] I typically issue one page full of unnamed equations.

Introductory Questionnaire

1. What do you hope to learn from this course?

2. What do you hope to do with this new knowledge?

3. What do you expect the lectures to do for you?

4. What do you expect the book to do for you?

5. How many hours do you think it will take to learn all you need to know from this course? Include everything: lectures, homework, etc.
    _____ hours/week

Figure 3.1 Introductory questionnaire. Courtesy: Prof. James Sethian, Department of Mathematics, University of California at Berkeley.

Questionnaire

1. What do you **love** about this class?

2. What do you **hate** about this class?

3. If you were teaching this class, what would **you** do?

4. If you could change one thing about this class, what would it be?

Figure 3.2   Second questionnaire, used four weeks into the semester. Courtesy: Prof. James Sethian, Department of Mathematics, University of California at Berkeley.

improve my teaching or to help the students? Second, are there still some expectations that are incompatible with what I will do? In the following lecture, I again make some time available to address these issues. The consequence of these questionnaires is that students become much more cooperative and willing to change their study habits.

Another important point is to make sure there is an atmosphere of cooperation in the classroom. Introductory science courses generally have the reputation of being extremely competitive. This is detrimental to *Peer Instruction* as competition is incompatible with collaboration. In my opinion, the best way to defuse such an atmosphere is to have an absolute grading scale. After analyzing grades for a couple of years, I found that averages tend to fluctuate very little from year to year. I therefore issue worksheets that allow students to track their progress and determine their final grade on an absolute scale.[2] I tell them that, while the average grade in previous years has hovered around a B–, there is no reason the average grade could not be an A—no student's grade will go down because others have done better.

[2] What I don't tell the students is that the grading scale I hand out tends to be a little tougher than the one I end up using. This disparity allows me a small amount of flexibility, and it significantly cuts down on the number of inquiries I get after issuing final grades.

The same atmosphere of cooperation is required for the convince-your-neighbors discussions. I therefore tell students that their performance on the *ConcepTests* will have no bearing on their final grade. I want them to be completely free of any pressure or competition. Although they are required to participate in the *ConcepTests*, they can, if they wish, participate anonymously.

The above principles set the right tone for *Peer Instruction*. Not all ideas may be applicable in any given setting, and by themselves these principles are not sufficient to make *Peer Instruction work*. Two points—preclass reading and examinations—need separate attention.

PRECLASS READING

At the first lecture, I distribute a schedule of reading assignments for the semester. And I stick to the schedule—better than I ever was able to do before. If a certain lecture goes faster than anticipated (a rare event), the students get an early break; nobody is unhappy. If a certain lecture goes more slowly than planned (usually because a *ConcepTest* revealed some difficulty with the material), I skip less important parts and rely on (*a*) the students' preclass reading, (*b*) weekly discussion sessions, and (*c*) homework assignments to cover these parts. In some cases, I may use part of the next lecture to stress some important points or to give an extra *ConcepTest*. In any case, I always schedule a review lecture in the middle of the semester to allow for some slack in an otherwise very rigid schedule. In short, the flexibility is in the schedule of each lecture, not in the semester schedule. This way I am able to cover the same amount of material as before.

Therefore, a key point is to get students to do part of the work ahead of the lecture. To make sure students carry out their reading assignments, I have implemented Just-in-Time Teaching, a very effective technique that is complementary to Peer Instruction and that is the subject of a separate book*. Before implementing Just-in-Time Teaching, I administered short three- or four-question reading quizzes at the beginning of each lecture period. From these quizzes, students earned a maximum of 10 bonus points toward their final grade, which was based on a 100-point scale. The bonus points reduced the weight of the final; for example, if a student earned 7 bonus points on the reading quizzes and if the final examination was worth a maximum of 40 points, that student's final examination score was rescaled by a factor 33/40 and the 7 bonus points added to the rescaled score. The overall effect of the bonus points is minor. The students, however, gladly used the opportunity to earn some extra points.

The reading quizzes were self-monitored. I usually allowed students to take the quiz anytime during a time interval from twenty minutes before class

*G. Novak, A. Gavrin, W. Christian, E. Patterson, *Just-in-Time Teaching: Blending Active Learning with Web Technology* (Prentice Hall, Upper Saddle River, NJ, 1999).

to five minutes after I had begun lecturing. To discourage collaboration on these quizzes, I graded them on a steep curve and renormalized all scores so that the average score was 6. This normalization ensured that the students did not help each other and that they would prevent others from abusing the system.

Although I strongly recommend using Just-in-Time-Teaching rather than reading quizzes, I am including a complete set of reading quizzes in Chapter 10 for those instructors who are not ready to implement Just-in-Time-Teaching. I have administered these quizzes in two ways: electronically (using *Classtalk*) and on paper forms. The paper forms can easily be graded using a template form.

EXAMINATIONS

Since 1992, I have included both conceptual essay questions and conventional problems on examinations. This mix is essential because exams determine the way in which many students study. In the words of John W. Moore, Professor of Chemistry at the University of Wisconsin-Madison: 'For the students, the exam is the dog and the course is the tail.' Mixed examinations are therefore the best way to make students aware of the increased emphasis on concepts.

At first sight, it may seem that conceptual questions make the examination easier, but as the example on dc circuits in Chapter 1 illustrates, the opposite is true for those who get by plugging numbers into equations (in my own case, half the students when I taught the class in a traditional fashion). Only those who understand the underlying physics consider conceptual questions straightforward.

A proper balance between computational and conceptual problems is important. I typically administer two midterm examinations and one final examination. On a midterm, four or five out of seven questions are conceptual; on the final, six out of twelve. Each problem carries the same weight because giving conceptual problems less weight would favor those who manage to solve problems without understanding what they are doing. At the beginning of the term, I issue a number of review exams pointing out the conceptual problems. The effect is to change the students' attitude right from the beginning.

4

A STEP-BY-STEP GUIDE TO PREPARING FOR A *PEER INSTRUCTION* LECTURE

In this chapter, I present a description of what I have done with my own material to change from a conventional lecturing style to *Peer Instruction*. I continue to use my old lecture notes and firmly believe that it is not necessary to completely rewrite one's lecture notes. I hope this description will therefore serve as a guide for converting your own material for use with *Peer Instruction*.

LECTURE OUTLINE

Each key point in a lecture takes a minimum of 15 minutes to cover: 7–10 minutes of lecturing, 5–8 minutes for a *ConcepTest*. One hour of lecturing therefore requires about four key points.

To convert a traditional lecture, I first decide which are the fundamental points that must be covered. In class, I no longer present definitions, derivations, and examples printed in the textbook or in the lecture notes. After taking these items out of my old presentation, I next determine the key points I want to get across. Eventually I am left with a skeletal lecture outline like the one shown in Figure 4.1.

1. Definition of pressure
2. Pressure as a function of depth
3. Archimedes' principle
4. Pascal's principle

Figure 4.1 Skeletal lecture outline for a lecture on fluid statics.

CONCEPTESTS

After making the outline, I select a number of conceptual questions for each key concept. Composing these questions from scratch constitutes perhaps the largest effort required to convert from a conventional lecture presentation to a *Peer Instruction* format. The importance of this task should not be underestimated, as the success of the method depends to a large extent on the quality and relevance of these questions.

While there are no hard-and-fast rules for the *ConcepTests*, they should at least satisfy a number of basic criteria. Specifically, they should

- focus on a single concept
- not be solvable by relying on equations
- have adequate multiple-choice answers
- be unambiguously worded
- be neither too easy nor too difficult

All these criteria directly affect feedback to the instructor. If more than one concept is involved in the question, it is difficult for the instructor to interpret the results and correctly gauge understanding. If students can derive the answer by relying on equations, the response does not necessarily reflect their real understanding. Ideally, the incorrect answer choices should reflect students' most common misconceptions. One can attempt to formulate the incorrect answers to each *ConcepTest* with this criterion in mind, but the ultimate source for the alternative responses (detractors) should be the students themselves. For instance, by posing the question in a fill-in format and then compiling the most frequent incorrect responses, a student-generated *ConcepTest* question that accurately mirrors common misconceptions is born. Examinations offer good opportunities for this.

The last two points are harder to gauge as a question is being created, even though they may sound entirely unmistakable. I have been surprised time and again to see questions that appeared to me completely straightforward and unambiguous be misinterpreted by many students. Needless to say, a question that is misinterpreted by students does not provide useful feedback. As for the level of difficulty, I made the point in Figure 2.3. Where a given question falls on this graph depends on the level and preparation of the students and on the clarity of the preclass reading material, the lecturing, and the question. Ideally, every teacher should make a similar plot to gauge the effectiveness of each question under his or her particular circumstances.

Chapter 11 contains *ConcepTest* questions covering most topics of an introductory physics course. The ones reproduced here have proved most effective in my own classes. Good sources of inspiration for additional *ConcepTests* are the end-of-chapter *questions* (as opposed to problems or exercises) in many standard introductory physics texts. In addition, the *Physics Teacher* and the *American Journal of Physics* publish many articles that may prove helpful in formulating new questions. Finally, books that emphasize fundamental concepts[1] often contain questions designed both to isolate these concepts and to help students grasp them by exposing common misconceptions.

To facilitate the exchange of questions and data, I recently created a World Wide Web server[2] where teachers around the world may submit and retrieve *ConcepTests*.

DEMONSTRATIONS

Lecture demonstrations can be used effectively in combination with *ConcepTests*, with one leading into the other. For instance, I can use a demonstration to lead into a question whose answer forces students to think about what they have just observed. Working the other way, I ask students to think about a particular question and use a demonstration to answer it. I have found that this combination greatly increases the value of demonstrations.

A good example is demonstrations on dc circuits. In my conventional lectures I used to omit most of these as the students didn't seem excited by seeing a light bulb's intensity change when a switch in a circuit is closed. In the new lecture format, I first give a *ConcepTest* such as the one shown in Figure 4.2. After everyone has answered the question the second time, I use the demonstration to show what happens. Because of the discussion, all students are now paying close attention, and when the switch is closed, one can hear the excitement of the students who got the right answer!

LECTURE

Peer Instruction lectures are much less rigid than those of the conventional method because, with the former, a certain amount of flexibility is necessary to respond to the sometimes unexpected results of the *ConcepTests*. I find myself improvising more often than before. While this may seem a disturbing prospect at first, the added flexibility actually makes the teaching *easier* than before. During the periods of silence (when the students are thinking), I get a break—a minute or so to catch my breath and reformulate my thoughts. During the convince-your-neighbors discussions, I participate in some of the discussions, as mentioned in Chapter 2. This participation benefits me in two ways. First, I hear students explain the answer in their own words. While my own explanations may be the most direct route from question to answer—the most efficient in terms of words and time—those provided by the students are often much more effective at convincing a skeptic, even if somewhat less direct. Sometimes the students

---

[1] See, *e.g.*: Arnold B. Arons, *A Guide to Introductory Physics Teaching*, New York: John Wiley & Sons, (1990); Paul G. Hewitt, *Conceptual Physics*, Boston: Scott Foresman, (1989); Jearl Walker, *The Flying Circus of Physics*, New York: John Wiley & Sons, (1977).

[2] This server is accessible via the uniform resource locator (URL) "http://galileo.harvard.edu" using Netscape or a similar Web client.

For identical light bulbs, when the switch is closed

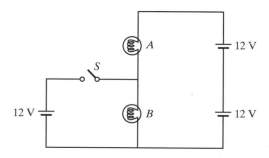

1. both bulbs go out.
2. the intensity of light bulb *A* increases.
3. the intensity of light bulb *A* decreases.
4. the intensity of light bulb *B* increases.
5. the intensity of light bulb *B* decreases.
6. some combination of 1–5.
7. nothing changes.

Figure 4.2 *ConcepTest* on dc circuits. For two identical light bulbs, the potential difference across each is 12 V, and so nothing happens when the switch is closed. Response statistics before (after) discussion: 1: 1% (0%), 2: 36% (42%), 3: 22% (7%), 4: 1% (3%), 5: 3% (2%), 6: 17% (7%), 7: 20% (39%).

offer a completely different perspective on the problem, in which case I frequently borrow from the students. In effect, they are teaching me how to teach. What is also important is that by listening to students who have reasoned their way to the wrong answer, I get a feel for what goes on in their minds. This involvement helps me to focus better on the problems they are facing and allows me to address these problems directly in my explanations. Finally, the personal interactions during the discussions help me keep in touch with the class.

The new lecture format elicits far more questions from the students than I have ever encountered before. Often these questions are very to the point and profound, and I usually attempt to address as many of them as I can.

EXAMINATIONS

Students are likely to ignore whatever changes one makes to the format or content of a class *unless these changes are reflected on the examinations*. For this reason, the examinations I give now contain both conceptual essay questions and standard textbook problems, as mentioned in Chapter 3. This is perhaps the best way to make the students aware of the increased emphasis on concepts.

Enrico Fermi was well known for his legendary ability to solve seemingly intractable problems—even in subjects entirely unfamiliar to him (*e.g.*, How many piano tuners in Chicago?). Such "Fermi problems" cannot be solved by deduction alone and require assumptions, models, order-of-magnitude estimates, and a great deal of self-confidence. Physicists in general frequently use back-of-the-envelope estimates to familiarize themselves with new problems. So why do we keep testing students with conventional problems? These problems contain the same number of unknowns and givens and frequently require nothing but mathematical skills. What distinguishes the successful scientist is not the ability to solve an integral, a differential equation, or a set of coupled equations but rather the ability to develop models, to make assumptions, to estimate magnitudes—just the skills developed in Fermi problems.

THE PROBLEM WITH CONVENTIONAL PROBLEMS

Let me use a simple example to illustrate what I believe is a serious problem with standard physics textbook exercises. I purposely choose an example outside the realm of physics for the following reason: Unless one has thought about this example, one is on equal footing with a student looking for the first time at a problem in a textbook. My example is based on a situation I encountered a while ago: I wanted to go shopping and pulled my car into a public parking lot near the stores. All spots were taken. Wanting to know if the best strategy was to roam around the lot or stay put in one spot, I decided to estimate the time I would have to wait if I stayed put. Using some rough estimates, I obtained a time of 3 minutes, and, sure enough, a space became available roughly 3 minutes later.

In Figure 4.3, this situation is posed as a classic Fermi problem, requiring students to (*a*) make assumptions, (*b*) make estimates, (*c*) develop a model, and (*d*) work that model out. Putting a question like this on an exam would surely cause a revolt among students. Let's therefore turn it into a typical textbook problem by removing, one-by-one, the requirements (*a*)–(*c*).

Because making assumptions typically is the last thing students are willing to do, let's start by making the assumption for them by adding a single sentence, as shown in Figure 4.4. In this form, the problem is still intractable to all but the smartest students because it presents an unfamiliar situation for which they have not yet developed (or seen) any model.

On a Saturday afternoon, you pull into a parking lot with unmetered spaces near a shopping area. You circle around, but there are no empty spots. You decide to wait at one end of the lot, where you can see (and command) about 20 spaces.

How long do you have to wait before someone frees up a space?

Figure 4.3   Fermi problem. This problem requires making assumptions and estimates, and developing and working out a model.

> On a Saturday afternoon, you pull into a parking lot with unmetered spaces near a
> shopping area. You circle around, but there are no empty spots. You decide to wait
> at one end of the lot, where you can see (and command) about 20 spaces. On average,
> people shop for about 2 hours.
>
> How long do you have to wait before someone frees up a space?

Figure 4.4   Rephrased version of the problem in Figure 4.3. The assumption that people shop
for about 2 hours has been added (a rough estimate but certainly in the right ballpark).

So, let's simplify even more by implicitly stating in the problem the result
one would get by statistically averaging over a large number of events, as in
Figure 4.5. In this form, the problem still would not fly because it presents an
unfamiliar situation and the model is not explicitly stated.

In Figure 4.6, I have finally turned the problem into a standard textbook
problem. At first glance, the statement is not much different from the initial
one in Figure 4.3, but the important thing is that somewhere in the book the students have seen (and subsequently highlighted and memorized) the equation:

$$t_{wait} = \frac{1}{2} \frac{T_{shop}}{N_{spaces}}$$

All that is left is for the students to plug in the numbers and use their calculators! In four steps, we have, so to say, thrown out the baby with the bath-
water. We have turned a question that requires a combination of skills relevant
for solving the type of problems scientists face into one that requires hardly
any skills at all. The original analytical challenges are now contained in the
equation and the problem statement. All opportunities to develop logical reasoning and to build confidence are lost.

WHY BOTHER?

The general idea behind this example is to the point: Most textbook problems
test mathematical instead of analytical thinking skills. Should this be cause for
concern? Even though we certainly manage to produce first-rate scientists with
the conventional way of teaching introductory science courses, I believe the an-

> On a Saturday afternoon, you pull into a parking lot with unmetered spaces near a
> shopping area. You circle around, but there are no empty spots. You decide to wait
> at one end of the lot, where you can see (and command) about 20 spaces. On average,
> people shop for about 2 hours.
>
> Assuming people leave at regularly-spaced intervals, how long do you have to wait
> before someone frees up a space?

Figure 4.5   Rephrased version of the problem in Figure 4.4. The assumption that people are
leaving the lot at regularly spaced intervals has been added.

> On a Saturday afternoon, you pull into a parking lot with unmetered spaces near a shopping area where people are known to shop, on average, for two hours. You circle around, but there are no empty spots. You decide to wait at one end of the lot, where you can see (and command) about 20 spaces.
>
> How long do you have to wait before someone frees up a space?

Figure 4.6 The Fermi question of Figure 4.4 turned into a standard textbook problem.

swer to this question is an emphatic "yes." I happen to believe that those who currently succeed in the sciences do so in spite of the current educational system, not because of it. I don't think we should be satisfied when a student just knows how to plug numbers into an equation in a given situation, how to solve a differential equation, or how to recite a law of physics. As the parking-lot problem makes clear, we need to look deeper than the standard textbook problem does.

Chapter 12 contains many class-tested conceptual exam questions. At first glance, some look simple, just like the light bulb problem in Chapter 1. If you are teaching an introductory physics course, you might include one or two of these questions on one of your examinations. Grading them will, I believe, reveal the true level of understanding your students have achieved. If the scores show a poor correlation between students' performance on conceptual and conventional problems, *Peer Instruction* can help improve the situation. For the dc-circuit problems of Figure 1.1, Figure 4.7 compares the scores I obtained in 1991 with those obtained in 1995 in a *Peer Instruction* setting. While the conventional problem score is close to its 1991 value,[3] note that *Peer Instruction* increases the average score for the conceptual problem by 70%.

[3] The 15% decrease in conventional problem score is larger than what I believe to be the grading accuracy (10%). Most likely the decrease occurred because in 1995 the students had done only one standard homework problem on dc circuits prior to the examination (more followed after the exam); in 1991 the students had completed an entire homework assignment on dc circuits.

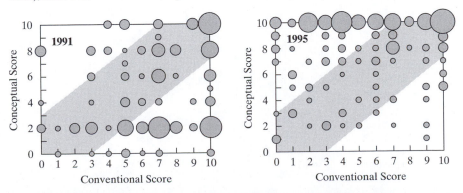

Figure 4.7 Correlation between scores on the conceptual and conventional examination problems from Figure 1.1. In 1991 the course was taught in a conventional manner; in 1995 the *Peer Instruction* method was used. The average scores in 1991 were 4.9 and 6.9 (out of 10) for the conceptual and conventional problems, respectively. In 1995 the average scores were 8.4 and 6.0, respectively.

5

SAMPLE LECTURE

As an example of *Peer Instruction*, let's consider a 90-minute lecture on Newton's laws, the outline of which is:

1. Newton's first law
2. Definitions of force and mass
3. Newton's second law
4. Newton's third law

Before coming to class, students are required to read the lecture notes as well as corresponding sections in the textbook. At the beginning of class, they complete the short reading quiz shown in Figure 5.1. Note that this quiz tests only whether or not the preclass reading was done; it does not test understanding of the material because doing so would penalize (and therefore discourage) the student who does the reading but is unable to master the concepts from the reading.

I use the same lecture notes I used when I taught this material conventionally. I describe the scope of classical mechanics and introduce Newton's first law by writing it on the chalkboard. After introducing the first law, I use a computer animation to show that it is really a statement about reference frames. Next, to firmly establish the relationship between forces and acceleration, I project the *ConcepTest* question shown in Figure 5.2. The students generally do well on this question, and its main purpose is to bolster their confidence. In any case, I don't dwell too long on this topic as Newton's other two laws generally cause far greater difficulties.

1. Which of these laws is not one of Newton's:
   1. To every action there is an opposed equal reaction.
   2. $F = ma$.
   3. All objects fall with equal acceleration.
   4. In the absence of a net external force, objects at rest stay at rest and objects in uniform motion stay in uniform motion.

2. The law of inertia
   1. is not covered in the reading assignment.
   2. expresses tendency of bodies to maintain their state of motion.
   3. is Newton's 3rd law.

3. "Impulse" is
   1. not covered in the reading assignment.
   2. another name for force.
   3. another name for acceleration.

Figure 5.1 Preclass reading quiz for lecture on particle dynamics. The correct answers are 1-3, 2-2, and 3-1. Response statistics: 1-1: 15%, 1-2: 2%, 1-3: 83%, 1-4: 0%, 2-1: 1%, 2-2: 98%, 2-3: 1%, 3-1: 82%, 3-2: 16%, 3-3: 2%. These and subsequent statistics in this chapter are from a representative semester during which *Peer Instruction* was used.

Then I define the concepts of force and mass and formulate Newton's second law. To make sure that the relationship between force, acceleration, and speed is clear, I use the question shown in Figure 5.3. The statistics in the caption show how the convince-your-neighbors discussion increases the number of correct responses and bolsters the students' confidence. With nearly 20% of the students providing wrong answers after the discussion, I would probably spend extra time discussing the correct answer.

A car rounds a curve while maintaining a constant speed.

Is there a net force on the car as it rounds the curve?

1. No—its speed is constant.
2. Yes.
3. It depends on the sharpness of the curve and the speed of the car.

Figure 5.2 *ConcepTest* on Newton's first law. Choice 2 is correct. Response statistics: 1: 3%, 2: 96%, 3: 1%.

A constant force is exerted on a cart that is initially at rest on an air track. Friction between the cart and the track is negligible. The force acts for a short time interval and gives the cart a certain final speed.

To reach the same final speed with a force that is only half as big, the force must be exerted on the cart for a time interval

1. four times as long as
2. twice as long as
3. equal to
4. half as long as
5. a quarter of

that for the stronger force.

Figure 5.3 *ConcepTest* on force. Choice 2 is correct. Response statistics before (after) discussion: 1: 16% (5%), 2: 65% (83%), 3: 19% (12%). Confidence before (after) discussion: pretty sure: 50% (71%); not quite sure: 43% (25%); just guessing: 7% (4%).

An important point in explaining this question is to avoid (at all cost!) using equations. My verbal argument goes as follows: Force causes acceleration, which tells how much an object's speed increases in a given interval of time. So if the force is half as large, the acceleration will be half as large. The force thus needs to act for a time interval twice as long to give the cart the same increase in speed.

The next *ConcepTest* (Figure 5.4) further elaborates on the previous question. Notice how much better the students do this time before the convince-your-neighbors discussion. With 90% providing the right answer before any discussion, there is little room for improvement. Still, the discussion does increase the students' confidence. The percentage of correct answers after discussion is a clear indication that not much further discussion of this question is required.

I immediately follow this *ConcepTest* with the one shown in Figure 5.5. To save time, I do not ask the students to discuss their answers.

With all these questions yielding more than 80% correct responses, I move on to Newton's third law, emphasizing that the two components of a third-law force pair never act on the same object. To make this point clear, I discuss the example of a person standing in an elevator. While the normal force exerted by the elevator floor on the person is equal to and opposite the weight of the person when the elevator is at rest, the two are not an action-reaction pair.

A constant force is exerted for a short time interval on a cart that is initially at rest on an air track. This force gives the cart a certain final speed.

The same force is exerted for the same length of time on another cart, also initially at rest, that has twice the mass of the first one. The final speed of the heavier cart is

1. one-fourth
2. four times
3. half
4. double
5. the same as

that of the lighter cart.

Figure 5.4 *ConcepTest* on force. Choice 3 is correct. Response statistics before (after) discussion: 1: 10% (1%), 3: 90% (99%). Confidence: pretty sure: 64% (95%); not quite sure: 34% (4%); just guessing: 2% (1%).

A constant force is exerted for a short time interval on a cart that is initially at rest on an air track. This force gives the cart a certain final speed.

Suppose we repeat the experiment but, instead of starting from rest, the cart is already moving in the direction of the force at the moment we begin to apply the force. After we exert the same constant force for the same short time interval, the increase in the cart's speed

1. is equal to two times its initial speed.
2. is equal to the square of its initial speed.
3. is equal to four times its initial speed.
4. is the same as when it started from rest.
5. cannot be determined from the information provided.

Figure 5.5 *ConcepTest* on force. Choice 4 is correct. Response statistics: 1: 10%, 2: 3%, 3: 5%, 4: 82%. Confidence: pretty sure: 63%; not quite sure: 35%; just guessing: 2%.

A locomotive pulls a series of wagons. Which is the correct analysis of the situation?

1. The train moves forward because the locomotive pulls forward slightly harder on the wagons than the wagons pull backward on the locomotive.

2. Because action always equals reaction, the locomotive cannot pull the wagons—the wagons pull backward just as hard as the locomotive pulls forward, so there is no motion.

3. The locomotive gets the wagons to move by giving them a tug during which the force on the wagons is momentarily greater than the force exerted by the wagons on the locomotive.

4. The locomotive's force on the wagons is as strong as the force of the wagons on the locomotive, but the frictional force on the locomotive is forward and large while the backward frictional force on the wagons is small.

5. The locomotive can pull the wagons forward only if it weighs more than the wagons.

Figure 5.6 *ConcepTest* on Newton's third law. Choice 4 is correct. Response statistics before (after) discussion: 1: 14% (7%), 2: 2% (2%) 4: 74% (86%), 5: 9% (5%). Confidence before (after) discussion: pretty sure: 59% (71%); not quite sure: 36% (26%); just guessing: 5% (3%).

When the elevator is accelerating, these two forces are no longer equal—the difference being responsible for accelerating the person. I make free-body diagrams for the person and the elevator and indicate which force pairs are third-law pairs. This presentation is followed by a lecture demonstration, immediately after which I confront the students with the classic question in Figure 5.6. In spite of the conceptual difficulty of this question, a surprisingly large fraction of the class answer correctly the first time around. This question always raises a large number of questions—it really gets students thinking—and I usually end up spending time after class explaining it a few more times.

Next, returning to the basic purpose of classical mechanics, I show the dual utility of Newton's second law: Given the forces on an object, one can use this law to determine the motion of that object. Conversely, by observing the motion of an object, one can use the law to deduce the forces exerted on that object. As examples, I cite the existence of the normal force, the forces on celestial bodies, and so forth.

I finally move on to the first force law—that of gravitation. I spend some time making clear the distinction between inertia (an object's tendency to maintain its state of motion) and gravitation (an object's tendency to attract matter): An astronaut on the Moon can easily lift a massive object, but kicking it would still hurt as much as it does on Earth.

The last question I use (Figure 5.7) involves gravitation but really tests the students' understanding of acceleration. This question offers the opportunity to spiral back and make the connection between the material in previous lectures (kinematics) and that in this lecture. While two thirds of the students provide the right answer, only a third are confident of their answer (the most frequent mistake is to assume that if speed increases, acceleration must increase too).

A cart on a roller-coaster rolls down the track shown below. As the cart rolls beyond the point shown, what happens to its speed and acceleration in the direction of motion?

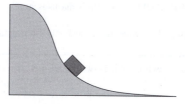

1. Both decrease.
2. The speed decreases, but the acceleration increases.
3. Both remain constant.
4. The speed increases, but acceleration decreases.
5. Both increase.
6. Other.

Figure 5.7 *ConcepTest* on gravitation, acceleration, and speed along an incline. Choice 4 is correct. Response statistics: 1: 3%, 2: 4%, 3: 8%, 4: 70%, 5: 11%, 6: 4%. Confidence: pretty sure: 34%; not quite sure: 57%; just guessing: 9%.

6

EPILOGUE

OFTEN-ASKED QUESTIONS

Whenever I explain the *Peer Instruction* method to other instructors, I invariably get asked many questions. Here, I address the most frequent.

Why bother?

The traditional measure of successful teaching has been the students' ability to solve problems. This tack assumes that one cannot solve a problem unless one understands the basics very well. As the example in Chapter 1 makes clear, however, a disturbingly large fraction of students develop strategies for solving problems without achieving even the most basic understanding of the concepts involved. So the question really is: Do we want our students to understand the basic principles, or are we satisfied if they can use formulas to solve numerical problems (even if close to half of them don't understand the underlying principles)? I am a clear advocate of teaching understanding, particularly so because my experience is that better understanding also leads to better problem solving.

Won't we be forced to cover less if the students spend time talking to each other during the lectures?

It is indeed not possible to cover the conventional amount of material in the lecture. I remedy this by putting a larger responsibility on the shoulders of the students. They read the material before coming to class, and I discuss only part of it in class. As a result, I still cover the standard material taught in courses around the nation.

Won't this new method drive an even bigger wedge between the students in an "Honors" section and those in a regular section?

As the graphs in Figs. 2.6 through 2.8 show, the gap between the bottom and the top end of the distribution is narrowed; thus, I believe it is safe to say that, while raising the overall performance of the class, *Peer Instruction* narrows the gap between Honors and regular students.

Peer Instruction may work for Harvard students, but it won't work for the students in my institution.

If you don't *expect* your students to read, they certainly won't, at Harvard or anywhere else. In most disciplines—the sciences are a notable exception—it would be unheard of to teach without asking students to read beforehand. I do not believe that students in the sciences are less literate than their colleagues in the humanities. The big problem—one that can be remedied—is that neither instructors nor students currently expect reading assignments in a science class. Books are supposed to elucidate unclear parts of the lectures and to provide practice problems, not introduce new material. In my opinion, science education requires an important change in attitude, one that cannot be accomplished overnight without some incentive. A simple incentive, which I have used with much success, is the reading quiz.

Isn't it accurate to say that students at institutions like Harvard are more articulate than students at less prestigious schools and therefore are better suited for interactive teaching?

At least 50 instructors have tried the method in a variety of academic settings: high schools, community colleges, large and small state schools. All have reported improvements in student performance and understanding.

I don't think I could use Peer Instruction. I don't have the right personality to do this sort of thing.

At first, the prospect of facing a classroom full of confusion may appear daunting. The *ConcepTests* often trigger many follow-up questions. The exact course of the lecture is unpredictable as it depends on the outcome of *ConcepTests*. However, those who have tried the method agree: Once one begins, it is impossible to go back to straight lecturing. First, the method provides an enormous amount of flexibility. Second, the discussions among (and with) students are extremely stimulating, independent of the personality of the instructor.

Does this way of lecturing constitute what we call "teaching"?

In a number of languages, the verb for "teaching" and "learning" is the same. Indeed, it is often expected that what is taught is learned. Unfortunately more often than not this is not the case. The goal of education should be to produce an environment where students can learn. *Peer Instruction* achieves this effectively and with relative ease.

How much work is required to change to the new lecture format?

Developing new *ConcepTests*, conceptual exam questions, and reading quizzes requires a significant effort. Once a set of these materials has been developed, however, teaching in the new format requires less preparation and effort than a traditional lecture because part of the lecture period is taken up by the student discussions. To help other instructors get started, I have included in this manual a complete set of classroom-tested *ConcepTests*, conceptual exam questions, and reading quizzes (see Chapters 10–12).

RECOMMENDATIONS

Having assisted a considerable number of people to adopt *Peer Instruction*, I recommend paying particular attention to the following five points.

1. *Convince yourself (and your colleagues).* The first thing to do is to convince yourself that changing the format of instruction makes sense. To this end, I recommend administering some sort of benchmark test, like the *Force Concept Inventory* (FCI) and *Mechanics Baseline* (MB) tests. Ideally, this should be done twice: once in a semester in which you use the conventional teaching method and once in a semester in which you use the *Peer Instruction* method. The same can be done with some conceptual exam questions. Data thus obtained will provide the clearest possible picture of what has been accomplished.

2. *Motivate students.* As I mentioned in Chapter 3, students resist change. Some will say, "Please teach us some *real* physics," by which they mean "please solve more problems for us" because this is what they have most likely been doing up to now. It is all too easy to succumb to this request. Motivating the students is half the work, and in the beginning of the semester, I spend considerable time and effort telling them what I will be doing during lectures and why I am doing it.

3. *Change examinations.* One of the best ways to make students accept a new method of teaching is to make the exams reflect the philosophy and goals of the course. About half the problems on the exams I give are conceptual. Although these questions are not unlike the questions I ask in class, they are usually longer and never multiple choice. I have provided a number of examples in Chapter 12. These questions may appear simple, but students generally find them harder than the standard test problems. I give these conceptual questions the same weight as the standard problems. The result is that, soon after having taken the first exam, students begin to appreciate the value of in-class questions. Typically, in the middle of the semester, some students will ask me to free up even more time for questions by abolishing lecturing altogether.

4. *Change lecture format.* As it is impossible to continue to present all the material in class *and* to devote time to in-class questions and discussions, I always urge people who decide to implement *Peer Instruction* to go all the way. Anything in between the traditional lecture and an interactive lecture is likely to produce nothing more than complaints from the students. If you continue to lecture as before, you are in effect telling the students they don't have to worry about the preclass reading because all the material will be treated in class anyway. In addition, there will be time only for a sporadic question or two during your traditional lecture, and the extra time taken by these questions will eventually cause you to fall behind schedule. This leads to a very undesirable situation: In addition to not doing their reading assignments, the students will start to see the in-class questions as a nuisance—as an addition to the traditional lecture rather than a substitute for it. The key is to rely on the students: If you make them read, they will. Reading assignments are therefore a must because students won't take their preclass reading seriously unless a reward is offered. The additional grading and bookkeeping more than pays for itself in the long term.

Another tactic I use to induce my students to come to class prepared is to pass out a reading assignment list at the beginning of the semester and then stress repeatedly that they are responsible for all material on this list *even if I do not cover it in class*. To reinforce this point, I issue problem set assignments on topics I skip, and I do not hesitate to set an exam problem on such topics.

5. *Problem solving.* Be sure to maintain adequate opportunities for the students to develop problem-solving abilities. It is possible to have students concentrate exclusively on conceptual problems at the expense of problem-solving, and to do so would be a mistake. An instructor who heard me talk on *Peer Instruction* adopted the method but completely abandoned problem solving, with the result that the class did poorly on a traditional examination. Opportunities for students to sharpen their problem-solving skills include homework assignments and problem-solving sessions. In my course, the homework assignments are worth 20% of the final grade; this ensures that the students take the assignments seriously.

PART TWO

RESOURCES

7

FORCE CONCEPT INVENTORY

On the following pages is the 1995 revision of the *Force Concept Inventory*[1]. The test should be given twice, once on the first day of class ("pretest") and once either at the end of or halfway through the semester ("posttest").

[1] The *Force Concept Inventory* (FCI) is a multiple-choice "test" designed to assess student understanding of the most basic concepts in Newtonian mechanics. The FCI can be used for several different purposes, but the most important one is to evaluate the effectiveness of instruction. For a full understanding of what has gone into the development of this instrument and of how it can be used, the FCI papers[a, b] should be consulted, as well as: (1) the papers on the *Mechanics Diagnostic Test*[c, d] the FCI predecessor, (2) the paper on the *Mechanics Baseline Test*[e] which is recommended as an FCI companion test for assessing quantitative problem solving skills, and (3) Richard Hake's[f] data collection on university and high school physics taught by many different teachers and methods across the USA.

[a] David Hestenes, Malcolm Wells, and Gregg Swackhamer. "Force Concept Inventory," *Phys. Teach.*, 30 (3), (1992), 141–151. Revised in 1995 by Ibrahim Halloun, Richard Hake, Eugene Mosca, and David Hestenes.

[b] David Hestenes and Ibrahim Halloun, "Interpreting the Force Concept Inventory," *Phys. Teach.*, 33 (8), (1995), 502, 504–506.

[c] Ibrahim Halloun and David Hestenes, "The Initial Knowledge State of College Physics Students," *Am. J. Phys.* 53 (11), (1985), 1043–1055.

[d] Ibrahim Halloun and David Hestenes. "Common Sense Concepts about Motion," *Am. J. Phys.*, 53 (11), (1985), 1056–1065.

[e] David Hestenes and Malcolm Wells, "A Mechanics Baseline Test," *Phys. Teach.*, 30 (3), (1992), 159–166.

[f] Richard Hake, "Survey of Test Data for Introductory Mechanics Courses," *AAPT Announcer,* 24 (2), (1994), 55; "Interactive-Engagement vs. Traditional Methods: A Six-Thousand-Student Survey of Mechanics Test Data for Introductory Physics Courses," preprint, June, 1995.

I tell students that the pretest is just for my own information—to determine their level of preparation (I call it a background test) so that I can adjust my first few lectures. I explicitly tell them the test will not affect their final grade. To get as many students as possible to take the test, I make it a requirement for enrolling in the class.

I typically schedule the posttest about a week before the midterm examination and announce it a week or so ahead of time. I tell the class they will get credit for this test—it counts like a regular homework assignment—and that it is the best way for them to test their own knowledge for the upcoming midterm. Until the test is taken, the students do not know it is the same test as the pretest.

Average nationwide scores run 25% to 70% for the pretest and 40% to 85% for the posttest. A score of 87% is considered "mastery"; 60% is the threshold for understanding Newtonian mechanics. An interesting quantity to determine is the fraction of the maximum possible gain realized[2]

$$G = \frac{S_f - S_i}{100 - S_i}$$

where S_i and S_f are the pre- and posttest scores in percent. For traditionally taught classes, $G \approx 0.25$; for classes that are taught in a more interactive manner, $0.36 < G < 0.68$.

The test is available on the enclosed diskette. You may print the test and use printed copies to evaluate your students' learning, provided you safeguard the integrity of the test as a diagnostic instrument. Since the test is used at many institutions it is important to prevent student circulation of the questions or the answer key. The authors suggest not referring to the test by the name "Force Concept Inventory" so as to shield the original literature from the students. Instead, a more generic name such as "Diagnostic Test" may be used. It is also advisable not to put too much weight on the score for the posttest. The test may not be otherwise distributed or edited. Answers are on page 58.

---

[2] R.R. Hake, "Survey of Test Data for Introductory Mechanics Courses," *AAPT Announcer,* 24 (2), (1994), 55; "Interactive-Engagement vs. Traditional Methods: A Six-Thousand-Student Survey of Mechanics Test Data for Introductory Physics Courses," preprint, June, 1995.

FORCE CONCEPT INVENTORY

1. Two metal balls are the same size but one weighs twice as much as the other. The balls are dropped from the roof of a single story building at the same instant. The time it takes the balls to reach the ground below will be

   1. about half as long for the heavier ball as for the lighter one.
   2. about half as long for the lighter ball as for the heavier one.
   3. about the same for both balls.
   4. considerably less for the heavier ball, but not necessarily half as long.
   5. considerably less for the lighter ball, but not necessarily half as long.

2. The two metal balls of the previous problem roll off a horizontal table with the same speed. In this situation

   1. both balls hit the floor at approximately the same horizontal distance from the base of the table.
   2. the heavier ball hits the floor at about half the horizontal distance from the base of the table than does the lighter ball.
   3. the lighter ball hits the floor at about half the horizontal distance from the base of the table than does the heavier ball.
   4. the heavier ball hits the floor considerably closer to the base of the table than the lighter ball, but not necessarily at half the horizontal distance.
   5. the lighter ball hits the floor considerably closer to the base of the table than the heavier ball, but not necessarily at half the horizontal distance.

3. A stone dropped from the roof of a single-story building to the surface of Earth

   1. reaches a maximum speed quite soon after release and then falls at a constant speed thereafter.
   2. speeds up as it falls because the gravitational attraction gets considerably stronger as the stone gets closer to Earth.
   3. speeds up because of an almost constant force of gravity acting upon it.
   4. falls because of the natural tendency of all objects to rest on the surface of Earth.
   5. falls because of the combined effects of the force of gravity pushing it downward and the force of the air pushing it downward.

4. A large truck collides head-on with a small compact car. During the collision

1. the truck exerts a greater amount of force on the car than the car exerts on the truck.

2. the car exerts a greater amount of force on the truck than the truck exerts on the car.

3. neither exerts a force on the other, the car gets smashed simply because it gets in the way of the truck.

4. the truck exerts a force on the car but the car does not exert a force on the truck.

5. the truck exerts the same amount of force on the car as the car exerts on the truck.

Use the statement and figure below to answer the next two questions (5 and 6).

The accompanying figure shows a frictionless channel in the shape of a segment of a circle with its center at *O*. The channel has been anchored to a frictionless horizontal table top. You are looking down at the table. Forces exerted by the air are negligible. A ball is shot at high speed into the channel at *P* and exits at *R*.

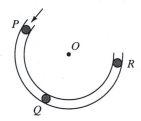

5. Consider the following distinct forces:

        A. a downward force of gravity.

        B. a force exerted by the channel pointing from *Q* to *O*.

        C. a force in the direction of motion.

        D. a force pointing from *O* to *Q*.

Which of the above forces is (are) acting on the ball when it is within the frictionless channel at position *Q*?

1. A only.

2. A and B.

3. A and C.

4. A, B, and C.

5. A, C, and D.

6. Which of the paths 1–5 below would the ball most closely follow after it exits the channel at *R* and moves across the frictionless table top?

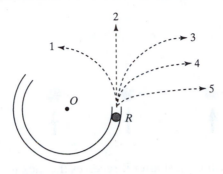

7. A steel ball is attached to a string and is swung in a circular path in a horizontal plane as illustrated in the figure below.

    At point *P*, the string suddenly breaks near the ball.

    If these events are observed from directly above, which of the paths 1–5 below would the ball most closely follow after the string breaks?

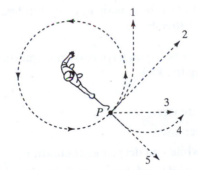

Use the statement and figure below to answer the next four questions (8–11).

The figure depicts a hockey puck sliding with constant speed v_0 in a straight line from point to *P* to point *Q* on a frictionless horizontal surface. Forces exerted by the air are negligible. You are looking down on the puck. When the puck reaches point *Q*, it receives a swift horizontal kick in the direction of the heavy print arrow. Had the puck been at rest at point *P*, then the kick would have set the puck in horizontal motion with a speed v_k in the direction of the kick.

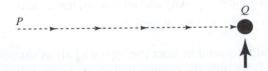

8. Which of the paths 1–5 below would the puck most closely follow after receiving the kick?

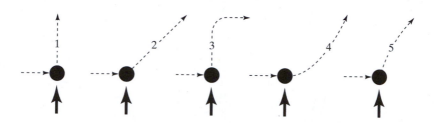

9. The speed of the puck just after it receives the kick is
  1. equal to the speed v_o it had before it received the kick.
  2. equal to the speed v_k resulting from the kick and independent of the speed v_o.
  3. equal to the arithmetic sum of the speeds v_o and v_k.
  4. smaller than either of the speeds v_o or v_k.
  5. greater than either of the speeds v_o or v_k, but less than the arithmetic sum of these two speeds.

10. Along the frictionless path you have chosen in question 8, the speed of the puck after receiving the kick
  1. is constant.
  2. continuously increases.
  3. continuously decreases.
  4. increases for a while and decreases thereafter.
  5. is constant for a while and decreases thereafter.

11. Along the frictionless path you have chosen in question 8, the main force(s) acting on the puck after receiving the kick is (are)
  1. a downward force of gravity.
  2. a downward force of gravity, and a horizontal force in the direction of motion.
  3. a downward force of gravity, an upward force exerted by the surface, and a horizontal force in the direction of motion.
  4. a downward force of gravity and an upward force exerted by the surface.
  5. none. (No forces act on the puck.)

12. A ball is fired by a cannon from the top of a cliff as shown below. Which of the paths 1–5 would the cannon ball most closely follow?

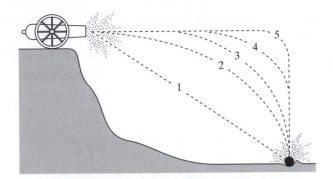

13. A boy throws a steel ball straight up. Consider the motion of the ball only after it has left the boy's hand but before it touches the ground, and assume that forces exerted by the air are negligible. For these conditions, the force(s) acting on the ball is (are)

   1. a downward force of gravity along with a steadily decreasing upward force.

   2. a steadily decreasing upward force from the moment it leaves the boy's hand until it reaches its highest point; on the way down there is a steadily increasing downward force of gravity as the ball gets closer to Earth.

   3. an almost constant downward force of gravity along with an upward force that steadily decreases until the ball reaches its highest point; on the way down there is only an almost constant downward force of gravity.

   4. an almost constant downward force of gravity only.

   5. none of the above. The ball falls back to ground because of its natural tendency to rest on the surface of the Earth.

14. A bowling ball accidentally falls out of the cargo bay of an airliner as it flies along in a horizontal direction.

   As observed by a person standing on the ground and viewing the plane as in the figure below, which of the paths 1–5 would the bowling ball most closely follow after leaving the airplane?

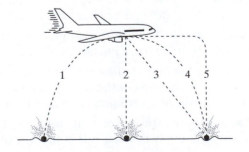

Use the statement and figure below to answer the next two questions (15 and 16).

A large truck breaks down out on the road and receives a push back into town by a small compact car as shown in the figure below.

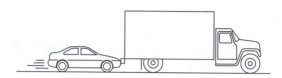

15. While the car, still pushing the truck, is speeding up to get up to cruising speed,

   1. the amount of force with which the car pushes on the truck is equal to that with which the truck pushes back on the car.

   2. the amount of force with which the car pushes on the truck is smaller than that with which the truck pushes back on the car.

   3. the amount of force with which the car pushes on the truck is greater than that with which the truck pushes back on the car.

   4. the car's engine is running so the car pushes against the truck, but the truck's engine is not running so the truck cannot push back against the car. The truck is pushed forward simply because it is in the way of the car.

   5. neither the car nor the truck exerts any force on the other. The truck is pushed forward simply because it is in the way of the car.

16. After the car reaches the constant cruising speed at which its driver wishes to push the truck,

   1. the amount of force with which the car pushes on the truck is equal to that with which the truck pushes back on the car.

   2. the amount of force with which the car pushes on the truck is smaller than that with which the truck pushes back on the car.

   3. the amount of force with which the car pushes on the truck is greater than that with which the truck pushes back on the car.

   4. the car's engine is running so the car pushes against the truck, but the truck's engine is not running so the truck cannot push back against the car. The truck is pushed forward simply because it is in the way of the car.

   5. neither the car nor the truck exerts any force on the other. The truck is pushed forward simply because it is in the way of the car.

17. An elevator is being lifted up an elevator shaft at a constant speed by a steel cable as shown in the following figure. All frictional effects are negligible. In this situation, forces on the elevator are such that

1. the upward force by the cable is greater than the downward force of gravity.
2. the upward force by the cable is equal to the downward force of gravity.
3. the upward force by the cable is smaller than the downward force of gravity.
4. the upward force by the cable is greater than the sum of the downward force of gravity and a downward force due to the air.
5. none of the above. (The elevator goes up because the cable is being shortened, not because an upward force is exerted on the elevator by the cable).

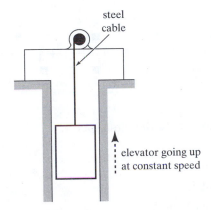

18. The following figure shows a boy swinging, starting at a point higher than *P*. Consider the following distinct forces:

    A. a downward force of gravity.
    B. a force exerted by the rope pointing from *P* to *O*.
    C. a force in the direction of the boy's motion.
    D. a force pointing from *O* to *P*.

Which of the above forces is (are) acting on the boy when he is at position *P*?

1. A only
2. A and B
3. A and C
4. A, B, and C
5. A, C, and D

19. The positions of two blocks at successive 0.20-s time intervals are represented by the numbered squares in the following figure. The blocks are moving toward the right.

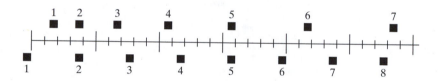

Do the blocks ever have the same speed?

1. No.
2. Yes, at instant 2.
3. Yes, at instant 5.
4. Yes, at instants 2 and 5.
5. Yes, at some time during the interval 3 to 4.

20. The positions of two blocks at successive 0.20-s time intervals are represented by the numbered squares in the figure below. The blocks are moving toward the right.

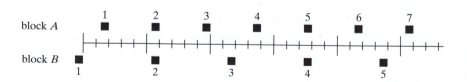

The accelerations of the blocks are related as follows:

1. The acceleration of *A* is greater than the acceleration of *B*.

2. The acceleration of *A* equals the acceleration of *B*. Both accelerations are greater than zero.

3. The acceleration of *B* is greater than the acceleration of *A*.

4. The acceleration of *A* equals the acceleration of *B*. Both accelerations are zero.

5. Not enough information is given to answer the question.

Use the statement and figure below to answer the next four questions (21 through 24).

A spaceship drifts sideways in outer space from point *P* to point *Q* as shown below. The spaceship is subject to no outside forces. Starting at position *Q*, the spaceship's engine is turned on and produces a constant thrust (force on the spaceship) at right angles to the line *PQ*. The constant thrust is maintained until the spaceship reaches a point *R* in space.

21. Which of the paths 1–5 below best represents the path of the spaceship between points *Q* and *R*?

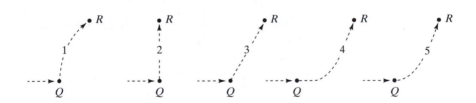

22. As the spaceship moves from point *Q* to point *R* its speed is
    1. constant.
    2. continuously increasing.
    3. continuously decreasing.
    4. increasing for a while and constant thereafter.
    5. constant for a while and decreasing thereafter.

23. At point *R*, the spaceship's engine is turned off and the thrust immediately drops to zero. Which of the paths 1–5 will the spaceship follow beyond point *R*?

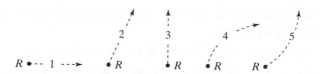

24. Beyond position R the speed of the spaceship is
    1. constant.
    2. continuously increasing.
    3. continuously decreasing.
    4. increasing for a while and constant thereafter.
    5. constant for a while and decreasing thereafter.

25. A woman exerts a constant horizontal force on a large box. As a result, the box moves across a horizontal floor at a constant speed v_0.
    The constant horizontal force applied by the woman
    1. has the same magnitude as the weight of the box.
    2. is greater than the weight of the box.
    3. has the same magnitude as the total force that resists the motion of the box.
    4. is greater than the total force that resists the motion of the box.
    5. is greater than either the weight of the box or the total force that resists its motion.

26. If the woman in the previous question doubles the constant horizontal force that she exerts on the box to push it on the same horizontal floor, the box then moves
    1. with a constant speed that is double the speed v_0 in the previous question.
    2. with a constant speed that is greater than the speed v_0 in the previous question, but not necessarily twice as great.
    3. for a while with a speed that is constant and greater than the speed v_0 in the previous question, then with a speed that increases thereafter.
    4. for a while with an increasing speed, then with a constant speed thereafter.
    5. with a continuously increasing speed.

27. If the woman in question 25 suddenly stops applying a horizontal force to the block, then the block
    1. immediately comes to a stop.
    2. continues moving at a constant speed for a while and then slows to a stop.

3. immediately starts slowing to a stop.

4. continues at a constant speed.

5. increases its speed for a while and then starts slowing to a stop.

28. In the following figure, student A has a mass of 75 kg and student B has a mass of 57 kg. They sit in identical office chairs facing each other.

Student A places his bare feet on the knees of student B, as shown. Student A then suddenly pushes outward with his feet, causing both chairs to move.

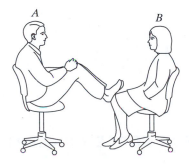

During the push and while the students are still touching one another,

1. neither student exerts a force on the other.

2. student A exerts a force on student B, but B does not exert any force on A.

3. each student exerts a force on the other, but B exerts the larger force.

4. each student exerts a force on the other, but A exerts the larger force.

5. each student exerts the same amount of force on the other.

29. An empty office chair is at rest on a floor. Consider the following forces:

       A. a downward force of gravity.

       B. an upward force exerted by the floor.

       C. a net downward force exerted by the air.

Which of the forces is (are) acting on the office chair?

1. A only

2. A and B

3. B and C

4. A, B, and C

5. None of the forces. (Since the chair is at rest, there are no forces acting upon it.)

30. Despite a very strong wind, a tennis player manages to hit a tennis ball with her racquet so that the ball passes over the net and lands in her opponent's court.

    Consider the following forces:

    A. a downward force of gravity.

    B. a force by the "hit."

    C. a force exerted by the air.

    Which of the above forces is (are) acting on the tennis ball after it has left contact with the racquet and before it touches the ground?

    1. A only

    2. A and B

    3. A and C

    4. B and C

    5. A, B, and C

ANSWER KEY FOR FORCE CONCEPT INVENTORY

Note: Question number is followed by correct answer.

1.	3	**11.**	4	**21.**	5
2.	1	**12.**	2	**22.**	2
3.	3	**13.**	4	**23.**	2
4.	5	**14.**	4	**24.**	1
5.	2	**15.**	1	**25.**	3
6.	2	**16.**	1	**26.**	5
7.	2	**17.**	2	**27.**	3
8.	2	**18.**	2	**28.**	5
9.	5	**19.**	5	**29.**	2
10.	1	**20.**	4	**30.**	3

8

MECHANICS
BASELINE TEST

On the following pages is the *Mechanics Baseline Test* developed by Hestenes and Wells.[1] Unlike the FCI test, this test requires the student to do a moderate amount of computation. Nationwide, the average score runs from 30% to 75%—typically about 15% below the average FCI posttest score.

This test should be given at the end of the semester. Many instructors give it as part of the final examination. I prefer to give it a week before the final examination, again telling the students that the test will count like a regular homework assignment and that it is the best way for them to test their knowledge for the upcoming final.

The test is available on the enclosed diskette. You may print the test and use printed copies to evaluate your students' learning, provided you safeguard the integrity of the test as a diagnostic instrument. Since the test is used at many institutions it is important to prevent student circulation of the questions or the answer key. The authors suggest not referring to the test by the name "Mechanics Baseline Test" so as to shield the original literature from the students. Instead, a more generic name such as "Diagnostic Test" may be used. It is also advisable not to put too much weight on the score for this test. The test may not be otherwise distributed or edited. Answers are on page 70.

[1]David Hestenes and Malcolm Wells, "A Mechanics Baseline Test," *Phys. Teach.*, 30 (3), (1992), 159–166.

MECHANICS BASELINE TEST

Refer to the figure below when answering the first two questions (1 and 2).

This diagram represents a multiflash photograph of an object moving along a horizontal surface. The positions indicated in the diagram are separated by equal time intervals. The first flash occurred just as the object started to move and the last just as it came to rest.

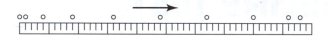

1. Which of the graphs 1–5 below best represents the object's velocity as a function of time?

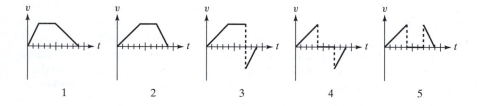

2. Which of the graphs 1–5 below best represents the object's acceleration as a function of time?

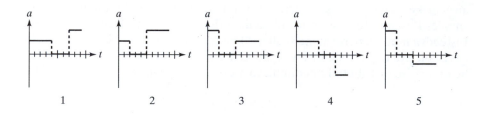

3. The velocity of an object as a function of time is shown in the following graph. Which of the graphs 1–5 best represents the net-force vs. time relationship for this object?

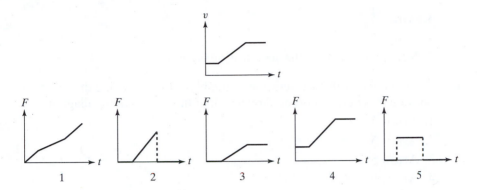

Refer to the following figure when answering the next three questions (4 through 6).

This diagram depicts a block sliding along a frictionless ramp. The eight numbered arrows in the diagram represent directions to be referred to when answering questions 4–6.

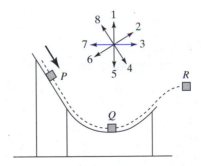

4. The direction of the acceleration of the block, when in position P, is best represented by which of the arrows in the diagram?

    1. Arrow 1
    2. Arrow 2
    2. Arrow 4
    4. Arrow 5
    5. None of the arrows, the acceleration is zero.

5. The direction of the acceleration of the block when in position Q is best represented by which of the arrows in the diagram?

    1. Arrow 1
    2. Arrow 3

3. Arrow 5

4. Arrow 7

5. None of the arrows, the acceleration is zero.

6. The direction of the acceleration of the block (after leaving the ramp) at position R is best represented by which of the arrows in the diagram?

1. Arrow 2

2. Arrow 3

3. Arrow 5

4. Arrow 6

5. None of the arrows, the acceleration is zero.

7. A person pulls a block across a rough horizontal surface at a *constant speed* by applying a force F. The arrows in the diagram below correctly indicate the directions, but not necessarily the magnitudes of the various forces on the block. Which of the following relations among the force magnitudes W, k, N, and F *must be true*?

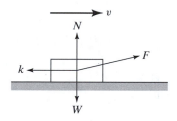

1. $F = k$ and $N = W$

2. $F = k$ and $N > W$

3. $F > k$ and $N < W$

4. $F < k$ and $N = W$

5. None of the above choices

8. A small metal cylinder rests on a circular turntable, rotating at a constant speed as illustrated in the diagram below. Which of the sets of vectors 1–5 below best describes the velocity, acceleration, and net force acting on the cylinder at the point indicated in the diagram?

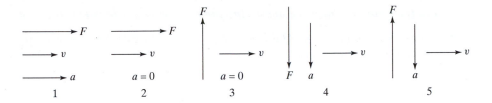

9. Suppose that the metal cylinder in the last problem has a mass of 0.10 kg
    and that the coefficient of static friction between the surface and the cylin-
    der is 0.12 If the cylinder is 0.20 m from the center of the turntable, at
    what maximum speed v can the cylinder move along its circular path with-
    out slipping off the turntable?

    1. $0 < v \leq 0.5$ m/s.
    2. $0.5 < v \leq 1.0$ m/s.
    3. $1.0 < v \leq 1.5$ m/s.
    4. $1.5 < v \leq 2.0$ m/s.
    5. $2.0 < v \leq 2.5$ m/s.

10. A young girl wishes to select one of the *frictionless* playground slides il-
    lustrated below to give her the greatest possible speed when she reaches
    the bottom of the slide.

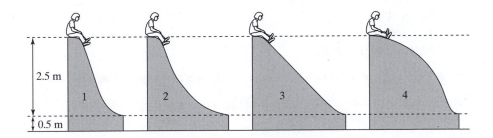

    Which of the slides illustrated in the diagram above should she choose?
    1. Slide 1
    2. Slide 2
    3. Slide 3
    4. Slide 4
    5. It doesn't matter, her speed would be the same for each slide.

Refer to the figure below when answering the next two questions (11 and 12).

P and *R* mark the highest and *Q* the lowest positions of a 50.0-kg boy swinging as illustrated in the following figure.

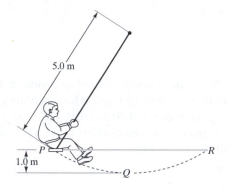

5.0 m

1.0 m

P

R

Q

11. What is the boy's speed at point *Q*?
   1. 2.5 m/s
   2. 7.5 m/s
   3. 10.0 m/s
   4. 12.5 m/s
   5. None of the above.

12. What is the tension in the rope at point *Q*?
   1. 250 N
   2. 525 N
   3. 7×10^2 N
   4. 1.1×10^3 N
   5. None of the above.

Refer to the figure below when answering the next two questions (13 and 14).

Blocks *A* and *B*, each with a mass of 1.0 kg, are hung from the ceiling of an elevator by ropes 1 and 2.

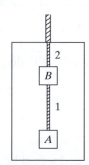

2

B

1

A

13. What is the force exerted by rope 1 on block *A* when the elevator is traveling upward at a constant speed of 2.0 m/s?

1. 2 N

2. 10 N

3. 12 N

4. 20 N

5. 22 N

14. What is the force exerted by rope 1 on block *B* when the elevator is stationary?

1. 2 N

2. 10 N

3. 12 N

4. 20 N

5. 22 N

Refer to the following figure when answering the next two questions (15 and 16).

The figure below depicts the paths of two colliding steel balls, *A* and *B*.

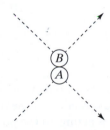

15. Which set of arrows best represents the direction of the change in momentum of each ball?

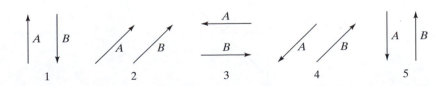

16. Which arrow best represents the impulse applied to ball B by ball A during the collision?

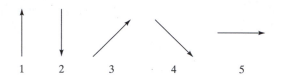

17. A car has a maximum acceleration of 3.0 m/s^2. What would its maximum acceleration be while towing a second car twice its mass?

1. 2.5 m/s^2

2. 2.0 m/s^2

3. 1.5 m/s^2

4. 1.0 m/s^2

5. 0.5 m/s^2

18. A woman weighing 6.0×10^2 N is riding an elevator from the 1st to the 6th floor. As the elevator approaches the 6th floor, it decreases its upward speed from 8.0 to 2.0 m/s in 3.0 s. What is the average force exerted by the elevator floor on the woman during this 3.0-s interval?

1. 120 N

2. 480 N

3. 600 N

4. 720 N

5. 1200 N

19. The diagram below depicts a hockey puck moving across a *horizontal, frictionless* surface in the direction of the dashed arrow. A constant force F, shown in the diagram, is acting on the puck. For the puck to experience a net force *in the direction of the dashed arrow,* another force must be acting in which of the directions 1–5 below?

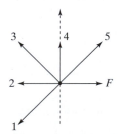

Refer to the following figure when answering the next three questions (20 through 22).

The diagram below depicts two pucks on a frictionless table. Puck *B* is four times as massive as puck *A*. Starting from rest, the pucks are pushed across the table by two *equal* forces.

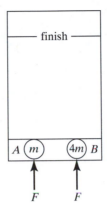

20. Which puck has the greater kinetic energy upon reaching the finish line?

1. Puck *A*
2. Puck *B*
3. They both have the same amount of kinetic energy.
4. too little information to answer

21. Which puck reaches the finish line first?

1. Puck *A*
2. Puck *B*
3. They both reach the finish line at the same time.
4. too little information to answer

22. Which puck has the greater momentum upon reaching the finish line?

1. Puck *A*
2. Puck *B*
3. They both have the same momentum.
4. too little information to answer

Refer to the following figure when answering the next three questions (23 through 25).

The graph below represents the motion of an object moving in one dimension.

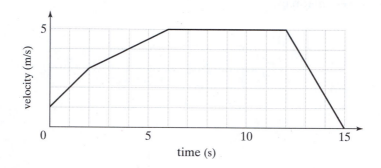

23. What was the object's average acceleration between $t = 0$ s and $t = 6.0$ s?
    1. 3.0 m/s^2
    2. 1.5 m/s^2
    3. 0.83 m/s^2
    4. 0.67 m/s^2
    5. none of the above

24. How far did the object travel between $t = 0$ and $t = 6.0$ s?
    1. 20.0 m
    2. 8.0 m
    3. 6.0 m
    4. 1.5 m
    5. none of the above

25. What was the average speed of the object for the first 6.0 s?
    1. 3.3 m/s
    2. 3.0 m/s
    3. 1.8 m/s
    4. 1.3 m/s
    5. none of the above

26. The figure below represents a multiflash photograph of a small ball being shot straight up by a spring. The spring, with the ball atop, was initially compressed to the point marked *P* and released. The ball left the spring at the point marked *Q*, and reached its highest point at the point marked *R*.

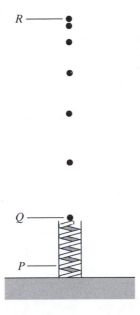

Assuming that the air resistance was negligible:

1. The acceleration of the ball was greatest just before it reached point *Q* (still in contact with the spring).

2. The acceleration of the ball was decreasing on its way from point *Q* to point *R*.

3. The acceleration of the ball was zero at point *R*.

4. All of the above responses are correct.

5. The acceleration of the ball was the same for all points in its trajectory from points *Q* to *R*.

ANSWER KEY FOR MECHANICS BASELINE TEST

Note: Question number is followed by correct answer.

1.	2	**10.**	5	**19.**	3
2.	4	**11.**	5	**20.**	3
3.	5	**12.**	3	**21.**	1
4.	3	**13.**	2	**22.**	2
5.	1	**14.**	2	**23.**	4
6.	3	**15.**	5	**24.**	1
7.	3	**16.**	1	**25.**	1
8.	4	**17.**	4	**26.**	5
9.	1	**18.**	2		

9

QUESTIONNAIRE RESULTS

This chapter presents the text of a handout which I issue after compiling the student responses to the introductory questionnaire completed by the students, which is shown on pp. 20–21. These responses are typical and vary little from year to year. I normally issue the handout in the second lecture and spend a few minutes discussing it, emphasizing the role of lectures and textbook in particular.

The text of this handout is available on the enclosed diskette and may be modified to suit your needs.

INTRODUCTORY QUESTIONNAIRE RESULTS

1. What do you hope to learn from this course?

I am excited to see that many of you hope to learn exactly what I am planning to teach! For those of you who were hoping for something else, however, I should clarify my goals. I plan to teach the basic principles of physics (to be specific, kinematics, conservation laws, mechanics, fluids, waves, and optics) and in the process:

1. teach you what physics is about;
2. provide an opportunity for you to sharpen your analytical thinking skills;
3. stimulate your curiosity and inquisitiveness about the world, provoke your questions, make you challenge conventional thoughts.

While there will be many topics in physics I will not have time to cover, I hope to give you the skills that will enable you to understand these topics as you encounter them in your future studies.

2.   What do you hope to do with this new knowledge?

Many interesting answers were given. I was gratified to see that many people replied that they hoped to use the knowledge and skills gained in this course in their own fields of interest. This is precisely what I hope to achieve in this course. I want the material we cover to be useful to you beyond the exam. I want you to become good critical and analytical thinkers, able to tackle not just familiar problems but also unknown new problems or questions. Not only to plug numbers into equations but able to develop new models and theories, to make qualified assumptions, and then use those models and assumptions to break new ground in science and technology.

3.   What do you expect the lectures to do for you?

There were many very thoughtful responses to this question, but I did encounter a number of misunderstandings about the lectures that I should address to avoid falling short of your expectations. The most serious misconception I encountered is that the lectures will present and explain the fundamental concepts, while the book will clarify the ideas presented in the lecture. This is *not* what is going to happen. You will be reading the material *before* coming to class. The book will introduce the basic terminology and definitions, hopefully raise some questions, perhaps even confuse you a little ("to wonder is to begin to understand"). The lectures are intended to challenge your thinking and thereby help you assess your understanding of the concepts you read about, to further and deepen your understanding of these concepts, to stimulate and inspire you, and to show you how things "fit together." The book will then provide further reference. In addition it will be a source for questions and problems.

Some of you expect to practice problem-solving in lecture, but problem-solving is not the main focus of this class. I want you to understand things, not just able to "plug and chug." This is clearly reflected in the way you will be tested—take a good look at the exams in the back of the syllabus. Close to half of the questions on each exam are not the traditional, quantitative problems you may have seen before. The solutions to many of these don't involve even a single equation. Rest assured, the sections and homework assignments will offer ample opportunity to sharpen your traditional problem-solving skills. The lectures are meant to stimulate your thinking, to further your basic understanding. I guarantee that a better understanding of the concepts will improve your problem-solving abilities, whereas the reverse is not necessarily true.

Here is what I think of some other answers given:

The purpose of the lectures:

is to	*is* not *to*
Be inspiring/stimulating	Cover all you are expected to know
Clarify the book	Practice problems
Explain confusing issues	Introduce the material
Make you think critically	
Give you lots to think about	
Spark further interest in the material	

4. *What do you expect the book to do for you?*

The following answer best reflects what I have in mind for both the book and the lectures: "I expect the book to be a kind of basic instruction manual, detailing the topics to be discussed, while the lectures will hopefully take this material and analyze it in a way so that I will be able to look at *any* problem intelligently, not just those from the book or the test."

Here is what I think of some of the answers given:

The purpose of the book:

is to	*is* not *to*
Provide background for lectures	Clarify the lectures
Be a resource for detailed explanations	
Be a reference and study guide	
Offer practice problems	
Teach problem solving	

5. *How many hours a week do you think it will take to learn all you need to know from this course?*

If you include lectures and sections, the answer should be somewhere in the 10–15 hrs/week range. By putting in any lesser amount of time, you will diminish your prospects of earning a good grade.

10

READING QUIZZES

As mentioned in Chapter 3, I currently strongly encourage the use of Just-in-Time Teaching* to get students to complete their reading assignments. If you are not ready to do so, the following pages contain the preclass reading quizzes for an introductory physics course, organized by subject. Note that most of the questions test knowledge of definitions rather than understanding. The answer key is on page 104.

All the quizzes in this chapter are available on the enclosed CD-ROM. You may cut, paste and print the quizzes to suit your classroom needs. The quizzes may not otherwise be distributed or edited without written permission from the publisher.

---

*G. Novak, A. Gavrin, W. Christian, E. Patterson, *Just-in-Time Teaching: Blending Active Learning with Web Technology* (Prentice Hall, Upper saddle River, NJ, 1999).

KINEMATICS

1. The slope of the curve in the position vs. time graph for a particle's motion gives
    1. the particle's speed.
    2. the particle's acceleration.
    3. the particle's average velocity.
    4. the particle's instantaneous velocity.
    5. not covered in the reading assignment

2. Is it possible for an object's instantaneous velocity and instantaneous acceleration to be of opposite sign at some instant of time?
    1. yes
    2. no
    3. need more information

3. Without air resistance, an object dropped from a plane flying at constant speed in a straight line will
    1. quickly lag behind the plane.
    2. remain vertically under the plane.
    3. move ahead of the plane.
    4. not covered in the reading assignment

4. A ball is thrown downward (not dropped) from the top of a tower. After being released, its downward acceleration will be
    1. greater than g.
    2. exactly g.
    3. smaller than g.
    4. not covered in the reading assignment

NEWTON'S LAWS

1. Which of these laws is not one of Newton's laws?
    1. Action is reaction.
    2. $F = ma$.
    3. All objects fall with equal acceleration.
    4. Objects at rest stay at rest, etc.

2. The law of inertia
    1. is not covered in the reading assignment.
    2. expresses the tendency of bodies to maintain their state of motion.
    3. is Newton's 3rd law.

3. "Impulse" is
    1. not covered in the reading assignment.
    2. another name for force.
    3. another name for acceleration.

FORCES

1. Viscous friction is
    1. larger than kinetic friction.
    2. equal to kinetic friction.
    3. smaller than kinetic friction.
    4. not covered in the reading assignment.

2. Astronauts on the Moon can jump so high because
    1. they weigh less there than they do on Earth.
    2. their mass is less there than it is on Earth.
    3. there is no atmosphere on the Moon.

3. Is the normal force on a body always equal to its weight?
    1. yes
    2. no
    3. not covered in the reading assignment

WORK

1. A woman holds a bowling ball in a fixed position. The work she does on the ball
    1. depends on the weight of the ball.
    2. cannot be calculated without more information.
    3. is equal to zero.

2. A man pushes a very heavy load across a horizontal floor. The work done by gravity on the load
   1. depends on the weight of the load.
   2. cannot be calculated without more information.
   3. is equal to zero.

3. When you do positive work on a particle, its kinetic energy
   1. increases.
   2. decreases.
   3. remains the same.
   4. need more information about the way the work was done

4. In a collision between two billiard balls,
   1. energy is not conserved if the collision is perfectly elastic.
   2. momentum is not conserved if the collision is inelastic.
   3. not covered in the reading assignment

CONSERVATIVE FORCES

1. The gravitational potential energy of a particle at a height z above Earth's surface
   1. depends on the height z.
   2. depends on the path taken to bring the particle to z.
   3. both 1 and 2.
   4. is not covered in the reading assignment.

2. Which of the following is not a conservative force?
   1. the force exerted by a spring on a particle in one dimension
   2. the force of friction
   3. the force of gravity
   4. not covered in the reading assignment

3. Which of the following was not discussed in the reading assignment?
   1. conservation of mechanical energy
   2. block and tackle
   3. power
   4. none of the above

POTENTIAL ENERGY

1. Suppose you know the potential energy function corresponding to a force. Is it always possible to calculate the force?

   1. yes
   2. only if the force is nonconservative
   3. not covered in the reading assignment

2. The potential energy of a spring is

   1. proportional to the amount the spring is stretched.
   2. proportional to the square of the amount the spring is stretched.
   3. not covered in the reading assignment.

3. A car slows down as a result of air friction. Which is true?

   1. The car's kinetic energy decreases.
   2. Heat is generated.
   3. The energy of the car/road/air system is constant.
   4. all of the above
   5. none of the above

GRAVITATION

1. Which is true? The gravitational force between two particles

   1. can be shielded by the presence of an intervening mass.
   2. is inversely proportional to the distance between the particles.
   3. obeys the law of superposition.
   4. is independent of the distance between the particles.

2. The gravitational constant G is

   1. equal to g at the surface of Earth.
   2. different on the Moon than on Earth.
   3. obtained by measuring the speed of falling objects having different masses.
   4. none of the above

3. Which is one of Kepler's laws?

   1. The gravitational attraction of Earth and the Sun provides a centripetal acceleration explaining Earth's orbit.

   2. The gravitational and inertial masses of an object are equivalent.

   3. The radial line segment from the Sun to a planet sweeps out equal areas in equal time intervals.

4. Which term was not introduced in today's reading assignment?

   1. escape velocity

   2. perihelion

   3. gravitational mass

   4. Hubble's constant

MOMENTUM

1. Which is true? Conservation of the total momentum of a system

   1. holds only when mechanical energy is conserved.

   2. holds for any system.

   3. follows from Newton's second law.

   4. is equivalent to Newton's third law.

2. The center of mass of a rigid object of arbitrary shape

   1. is always inside the object.

   2. can lie outside the object.

   3. depends on the motion of the object.

   4. depends on the frame of reference of the object.

3. Compared with the kinetic energy of its center of mass (CM), the total kinetic energy of a system is

   1. always less than the kinetic energy of the CM.

   2. always equal to the kinetic energy of the CM.

   3. greater than or equal to the kinetic energy of the CM.

   4. depends on the particular system

4. A rocket is propelled forward by ejecting gas at high speed. The forward motion is a consequence of

   1. conservation of energy.

   2. conservation of momentum.

   3. both of the above.

   4. neither of the above.

COLLISIONS

1. The impulse delivered to a body by a force is
    1. defined only for interactions of short duration.
    2. equal to the change in momentum of the body.
    3. equal to the area under an F vs. x graph.
    4. defined only for elastic collisions.

2. In an elastic collision
    1. energy is conserved.
    2. momentum is conserved.
    3. the magnitude of the relative velocity is conserved.
    4. all of the above

3. In an inelastic collision
    1. both energy and momentum are conserved.
    2. energy is conserved.
    3. momentum is conserved.
    4. neither is conserved.

4. In two-dimensional elastic collisions, the conservation laws
    1. allow us to determine the final motion.
    2. place restrictions on possible final motions.
    3. do not allow us to say anything about the final motion.
    4. are not covered in the reading assignment.

ROTATIONAL KINEMATICS I

1. An object is rotated about a vertical axis by 90° and then about a horizontal axis by 180°. If we start over and perform the rotations in the reverse order, the orientation of the object
    1. will be the same as before.
    2. will be different than before.
    3. depends on the shape of the object.
    4. is not covered in the reading assignment.

2. A disk is rotating at a constant rate about a vertical axis through its center. Point Q is twice as far from the center of the disk as point P is. The angular velocity of Q at a given time is

1. twice as big as P's.
2. the same as P's.
3. half as big as P's.
4. none of the above.

3. When a disk rotates counterclockwise at a constant rate about a vertical axis through its center, the tangential acceleration of a point on the rim is
   1. positive.
   2. zero.
   3. negative.
   4. impossible to determine without more information.

ROTATIONAL KINEMATICS II

1. The rotational inertia of a rigid body
   1. is a measure of its resistance to changes in rotational motion.
   2. depends on the location of the axis of rotation.
   3. is large if most of the body's mass is far from the axis of rotation.
   4. is all of the above.
   5. is none of the above.

2. The angular momentum of a particle
   1. is independent of the specific origin of coordinates.
   2. is zero when its position and momentum vectors are parallel.
   3. is zero when its position and momentum vectors are perpendicular.
   4. is not covered in the reading assignment.

3. Which term was not introduced in today's reading assignment?
   1. axis of rotation
   2. rotational kinetic energy
   3. gyroscopes
   4. moment of inertia

ROTATIONAL DYNAMICS I

1. When a force F acts on a body, the perpendicular distance between the line of action of F and the origin of coordinates is called the
   1. torque.
   2. moment arm.

   3. angular momentum.

2. The equation of motion for a rotating body, $\tau = dL/dt$,

   1. is a new law of physics.

   2. can be derived from Newton's laws.

   3. can be derived, but depends on laws other than Newton's.

3. A wheel rolls without slipping along a horizontal surface. The center of the wheel has a translational speed v. The lowermost point on the wheel has a net forward velocity

   1. $2v$.

   2. v.

   3. zero.

   4. need more information

ROTATIONAL DYNAMICS II

1. The moment of inertia of a rigid body about a fixed axis through its center of mass is I. The moment of inertia of this same body about a parallel axis through some other point is always

   1. smaller than I.

   2. the same as I.

   3. larger than I.

   4. whether it's larger or smaller depends on the choice of axis

2. A disk rolls without slipping along a horizontal surface. The center of the disk has a translational speed v. The uppermost point on the disk has a translational speed

   1. 0.

   2. v.

   3. $2v$.

   4. need more information

3. An ice-skater spins about a vertical axis through her body with her arms held out. As she draws her arms in, her angular velocity

   1. increases.

   2. decreases.

   3. remains the same.

   4. need more information

OSCILLATIONS

1. The time interval for one repetition of the cycle in simple harmonic motion is called the
   1. frequency.
   2. period.
   3. amplitude.
   4. phase.

2. The frequency of a coupled mass-spring oscillator depends on
   1. the value of the spring constant alone.
   2. the value of the mass alone.
   3. both of the above
   4. neither of the above

3. The total energy of a frictionless mass-spring oscillator
   1. is constant.
   2. depends on the amplitude of the oscillations.
   3. both of the above
   4. is not covered in the reading assignment.

4. Which term is not associated with forced oscillations?
   1. sympathetic oscillation
   2. driving force
   3. Doppler shift
   4. resonance

WAVES

1. A transverse wave propagates along a string. The particles in the string move
   1. perpendicular to the direction of propagation.
   2. parallel to the direction of propagation.
   3. depends on the initial disturbance
   4. not covered in the reading assignment

2. The speed of a wave on a string depends on
   1. the amplitude of the wave.
   2. the material properties of the string.

   3. both of the above.

   4. neither of the above.

3. Beats occur when two superimposed waves are of
  1. slightly different amplitudes and the same frequency.
  2. slightly different frequencies.
  3. the opposite amplitude and identical frequency.
  4. the same amplitude and frequency, but different phase.

4. Antinodes and nodes occur
  1. during beats.
  2. in standing waves.
  3. in traveling waves.
  4. in longitudinal waves.
  5. in more than just one of the above.
  6. in none of the above.

SOUND

1. Which of the following characterize(s) sound waves in air?
  1. They are longitudinal.
  2. The restoring force is supplied by air pressure.
  3. The density of the air molecules oscillates in space.
  4. 1 and 2
  5. 1 and 3
  6. 1, 2, and 3

2. A standing sound wave in a tube having one open end has a displacement
  1. antinode at the closed end and node at the open end.
  2. antinode at the closed end and at the open end.
  3. node at the closed end and antinode at the open end.
  4. node at the closed end and at the open end.

3. You are at rest on a platform at a railroad station. A train approaches the platform blowing its whistle. As the train passes you, the pitch of the whistle
  1. increases.
  2. decreases.
  3. stays the same.
  4. depends on the amplitude of the sound.

4. Seismic waves differ from sound waves in that seismic waves
    1. have a restoring force provided by the elasticity of Earth.
    2. may propagate transversely.
    3. both of the above
    4. neither of the above

FLUID STATICS

1. Which statement does not apply? In the steady flow of an incompressible fluid,
    1. the flow velocity at a point is tangent to the streamline through that point.
    2. the density of the fluid is proportional to the density of streamlines.
    3. streamlines cannot cross each other.
    4. the wider the streamline spacing, the lower the velocity of the flow.

2. A fluid is
    1. a liquid.
    2. a gas.
    3. anything that flows.
    4. anything that can be made to change shape.

3. A static fluid in a container is subject to both atmospheric pressure at its surface and Earth's gravitation. The pressure at the bottom of the container
    1. depends on the height of the fluid column.
    2. depends on the shape of the container.
    3. is equal to the atmospheric pressure.

4. The buoyant force on an immersed body has the same magnitude as
    1. the weight of the body.
    2. the weight of the fluid displaced by the body.
    3. the difference between the weights of the body and the displaced fluid.
    4. the average pressure of the fluid times the surface area of the body.

FLUID DYNAMICS

1. The equation of continuity says that the velocity of fluid flow in a pipe is inversely proportional to the cross-sectional area
    1. only for an incompressible fluid.
    2. only for a horizontal pipe.

3. both of the above

4. always

2. Bernoulli's equation is a conservation law for

1. momentum.

2. energy.

3. mass.

4. streamlines.

3. Which situation cannot be described with Bernoulli's equation?

1. the flow of water out of a tank having a small hole near its bottom

2. the steady flow of water in a fire hose

3. the static pressure distribution due to the air velocities near (but not at) airfoil surfaces.

4. fluid flow through a pump equipped with a piston

4. When the velocity of a fluid flow increases, pressure decreases. This relationship is expressed by

1. Pascal's principle.

2. the equation of continuity.

3. Bernoulli's equation.

4. none of the above.

ELECTROSTATICS I

1. Which of the following is not true? The electric force

1. decreases with the inverse of the square of the distance between two charged particles.

2. between an electron and a proton is much stronger than the gravitational force between them.

3. between two protons separated by a distance d is larger than that between two electrons separated by the same distance d.

4. may be either attractive or repulsive.

2. A material that permits electric charge to move through it is called a(n)

1. insulator.

2. conductor.

3. capacitor.

4. inductor.

3. When the electric charge on each of two charged particles is doubled, the electric force between them is

    1. doubled.

    2. quadrupled.

    3. the same.

    4. none of the above

4. In any reaction involving charged particles, the total charge before and after the reaction is always the same. This relationship is known as

    1. quantization of charge.

    2. conservation of charge.

    3. the law of induction.

    4. not covered in the reading assignment

ELECTROSTATICS II

1. Which statement is not true?

    1. The electric field obeys the principle of superposition.

    2. The tangent to an electric field line at a point gives the direction of the field at that point.

    3. The density of electric field lines is directly proportional to the strength of the field.

    4. Negative charges are sources of electric field lines and positive charge sinks.

2. An electric dipole in a uniform electric field experiences

    1. only a net external force.

    2. only a torque.

    3. both a net external force and a torque.

    4. neither a net external force nor a torque.

    5. answer depends on the strength of the field

3. Which is (are) true?

    1. The electric flux through a closed surface whose volume holds a net charge Q depends on both Q and the surface area.

    2. For charges at rest, Coulomb's law and Gauss' law are equivalent.

    3. both 1 and 2

    4. neither 1 nor 2

4. Which is (are) true? When the charge distribution on a conductor reaches equilibrium,

    1. the electric field within the conductor is zero.
    2. any electric charge deposited on the conductor resides on the surface.
    3. the electric field at the surface is perpendicular to the surface.
    4. all of the above
    5. two of the above
    6. none of the above

ELECTRIC POTENTIAL I

1. A charge q is placed a distance r from the origin, and a charge $2q$ is placed a distance $2r$. There is a charge Q at the origin. If all charges are positive, which charge is at the higher potential?

    1. q
    2. $2q$
    3. The two charges have the same potential.

2. Which charge in question 1 has the higher electrostatic potential energy?

    1. q
    2. $2q$
    3. The two charges have the same potential energy.

3. A spherical metal shell carries a uniform positive surface charge. The potential is the same over the surface of the shell. Which statement is correct?

    1. The potential is highest at the geometrical center of the shell volume.
    2. The potential is lowest at the geometrical center of the shell volume.
    3. The potential at the center of the shell volume is the same as on the shell surface.

ELECTRIC POTENTIAL II

1. Which statement(s) is(are) true? The electric potential energy of a charge distribution is

    1. equal to the amount of work required to bring the charges to their final configuration if they are initially separated by large distances.
    2. proportional to the square of the electric field generated by the charges.

3. both of the above

4. neither of the above

2. The amount of energy required to assemble a point charge is called the charge's

1. capacitance.

2. self-energy.

3. field strength.

4. not covered in the reading assignment.

3. Two isolated metallic spheres each have a net charge Q uniformly distributed over their surfaces. One sphere has a radius r and the other has a radius R, where $R > r$. Which charge distribution stores more electric energy?

1. the sphere of radius r.

2. the sphere of radius R.

3. need more information.

CAPACITANCE

1. Two identical capacitors are connected first in parallel and then in series. Which combination has the greater capacitance?

1. the pair in parallel

2. the pair in series

3. the two combinations have the same capacitance

2. Which statement(s) is(are) true? A dipole moment is created in a dielectric placed in an electric field when

1. molecules or atoms of the dielectric material become polarized.

2. randomly oriented permanent dipoles in the material realign themselves.

3. both 1 and 2, with the particular mechanism depending on the material

4. none of the above.

3. Compared with the applied electric field, the electric field within a linear dielectric is

1. smaller.

2. larger.

3. depends on the dielectric

4. In order to increase the energy stored in a parallel-plate capacitor when an electric potential is applied, we should
  1. increase the area of the plates.
  2. increase the separation between the plates.
  3. insert a dielectric between the plates.
  4. all of the above
  5. two of the above
  5. none of the above

OHM'S LAW

1. Which statement(s) is(are) true? When a long straight conducting wire of constant cross-section is connected to the terminals of a battery, the electric field
  1. lines are uniformly distributed over the cross-sectional area of the conductor.
  2. inside the wire is of constant magnitude and its direction is parallel to the wire.
  3. both of the above
  4. neither of the above

2. Which statement(s) is(are) true? Ohm's law
  1. asserts that the current in a conducting wire is proportional to the resistance of the wire.
  2. is a general law of nature like Newton's laws and Gauss' law.
  3. describes the electrical properties of some conducting materials.
  4. all of the above
  5. two of the above

3. Which term was not defined in the reading assignment?
  1. drift velocity
  2. impedance
  3. superconductivity
  4. resistivity

4. Two identical resistors are connected first in series and then in parallel. Which combination has the larger net resistance?
  1. the pair in series
  2. the pair in parallel
  3. The two combinations have the same resistance.

DC CIRCUITS

1. Which is(are) true? The emf of a source of electric potential energy is

   1. the amount of electric energy delivered by the source per coulomb of positive charge as this charge passes through the source from the low- to the high-potential terminal.

   2. equal in magnitude to the potential drop in the external circuit connected between the terminals of the source of emf.

   3. both of the above

   4. neither of the above

2. Which is(are) true? Kirchhoff's second rule

   1. relates the sum of the emfs around a closed loop in a circuit to the potential changes across all resistors and circuit elements.

   2. implies conservation of energy in electric circuits.

   3. relates the currents entering and leaving any branch point in a circuit.

   4. all of the above

   5. two of the above

   6. none of the above

3. A Wheatstone bridge is a device used to measure

   1. current.

   2. potential.

   3. resistance.

   4. joule-heating losses.

4. A resistor and an initially uncharged capacitor arranged in series are charged by a battery, which is connected at $t = 0$. The current in the circuit

   1. is constant because the emf supplied by the battery is constant.

   2. decreases exponentially in time.

   3. increases exponentially in time.

   4. There is no current because the electrons cannot flow through the gap in the capacitor.

MAGNETOSTATICS

1. Two charges q and Q move with nonzero velocities with respect to a fixed reference frame. The magnetic force on q exerted by Q is

   1. perpendicular to the velocity of q and depends only on the velocity of Q.

2. perpendicular to the velocity of q and depends on both the velocity of Q and that of q.

3. perpendicular to the velocity of Q and depends only on the velocity of q.

4. perpendicular to the velocity of Q and depends on both the velocity of Q and that of q.

2. Which is(are) true?

1. The magnetic field lines of a moving charge form closed loops.

2. The magnetic field obeys the principle of superposition.

3. The magnetic flux through a closed surface is proportional to the total number of magnetic poles enclosed within the surface.

4. all of the above

5. two of the above

6. none of the above

3. A long straight wire lies along the x-axis and carries a current of electrons that move in the positive x-direction. The magnetic field due to this current, at a point P on the negative y-axis, points in which direction?

1. $+x$

2. $-x$

3. $+y$

4. $-y$

5. $+z$

6. $-z$

4. Which is(are) true? The magnetic dipole moment of a current loop

1. is proportional to the area enclosed by the loop.

2. is proportional to the current in the loop.

3. is well defined only when the observer is far from the loop.

4. all of the above

5. two of the above

6. none of the above

AMPÈRE'S LAW

1. Ampère's law gives the magnetic field produced by a distribution of currents. Which condition(s) must be satisfied?

1. The distribution of currents must be steady.

2. In order to solve, the distribution must have sufficient symmetry.

3. both of the above

4. neither of the above

2. Which is(are) true? The magnetic field inside a solenoid

1. is parallel to the axis of the solenoid.

2. has circular field lines centered on the axis.

3. has a magnitude that is proportional to the total number of turns.

4. all of the above

5. two of the above

HALL EFFECT

1. The Hall effect

1. provides empirical evidence that the charge carriers in metals are negative.

2. can be used to determine the density of free electrons in a metal.

3. both of the above

4. neither of the above

2. A small planar current loop is placed in a uniform magnetic field. The magnitude of the torque on the loop is a maximum when

1. the plane of the loop is parallel to the direction of the field.

2. the plane of the loop is perpendicular to the direction of the field.

3. the angle between the plane of the loop and the magnetic field is somewhere between 0 and 90°.

4. the torque is independent of the angle between its plane and the magnetic field

MAGNETIC INDUCTANCE

1. Which is true?

1. The field lines of an induced electric field form closed loops.

2. The induced electric field is conservative.

3. both of the above

4. neither of the above

2. The magnetic energy stored in an inductor is

1. proportional to the square of the current through the inductor.

2. proportional to the square of the magnetic field of the inductor.

3. both of the above

4. neither of the above

MUTUAL INDUCTANCE

1. Two current-carrying coils of wire are in close proximity. We can change the mutual inductance of the pair by

1. changing the relative positions of the coils.

2. changing the currents.

3. increasing the number of turns in one of the coils.

4. all of the above.

5. two of the above.

2. A resistor R and an inductor L are connected in series to a battery, which is switched on at $t = 0$. The current in the circuit is time-dependent. If we repeat the experiment with a resistor of resistance $5R$, the time constant

1. decreases by a factor of 5.

2. increases by a factor of 5.

3. does not change.

AC CIRCUITS I

1. In a circuit consisting of a resistor connected to an oscillating source of emf, the current

1. leads the emf.

2. lags behind the emf.

3. is in phase with the emf.

4. the answer depends on the source of emf

2. A capacitor is connected to an oscillating source of emf. As the frequency of the emf increases, the capacitive reactance

1. increases.

2. decreases.

3. remains the same.

4. depends on the direction of the current.

3. In a dc circuit (which means the frequency of the source of emf is zero), which circuit element presents the greatest "resistance" to charge flow?

1. capacitor

2. inductor

3. resistor

4. Answer depends on the relative values of $C, L,$ and R.

4. The current in an ac circuit is represented by a phasor. The value of the current at some time t is given by

1. the length of the phasor.

2. the value, in radians, of the angle between the phasor and the horizontal axis.

3. the projection of the phasor on the vertical axis.

4. the projection of the phasor on the horizontal axis.

AC CIRCUITS II

1. A capacitor having an initial charge Q and an inductor are connected in series. The energy in the inductor is a maximum when the charge on the capacitor is

1. Q.

2. $\frac{1}{2} Q$.

3. zero.

4. the energy does not depend on the charge

2. A capacitor having an initial charge Q is connected in series with an inductor and a resistor. As a function of time, the charge on the capacitor

1. oscillates sinusoidally.

2. oscillates sinusoidally with exponentially decreasing amplitude.

3. does not vary in time as there is no driving emf.

4. not covered in the reading assignment

3. Which of the following terms were introduced in the reading assignment to describe an RLC circuit having an external emf?

1. resonance

2. impedance

3. bandwidth

4. all of the above

4. In transmitting electricity from a power plant to the consumer, transformers are utilized for which of the following tasks?

1. stepping up the output voltage at the power plant

2. stepping down the voltage just before it reaches the consumer

3. both of the above

4. neither of the above

MAXWELL'S EQUATIONS

1. A capacitor has been charged to a constant potential V. The displacement current between its plates

    1. is equal to the current that was required to charge up the capacitor.

    2. depends on the Ampèrian surface chosen.

    3. is zero.

    4. induces a magnetic field.

2. The Maxwell modification of Ampère's law describing the creation of a magnetic field is the analog of

    1. Gauss' law on electric fields and charges.

    2. Gauss' law on magnetic fields and poles.

    3. the Lorentz equation.

    4. Faraday's law.

ELECTROMAGNETIC WAVES I

1. An electromagnetic wave polarized in the positive y direction propagates in the negative z-direction. What is the direction of the magnetic field?

    1. $+x$

    2. $-y$

    3. $-x$

    4. $+z$

2. In a planar harmonic wave, the magnetic field achieves its maximum when the electric field

    1. is also at its maximum.

    2. is at its minimum.

    3. is at some intermediate value.

    4. the relationship between electric and magnetic fields depends on the plane wave

ELECTROMAGNETIC WAVES II

1. Which is(are) true? The energy carried by an electromagnetic wave in a vacuum

    1. propagates at the speed of light.

    2. consists of equal contributions from the electric and magnetic fields.

    3. propagates along the direction of the electric field.

    4. all of the above

    5. two of the above

2. A grain of interplanetary dust is in the Sun's gravitational field. If we consider the grain to be isolated from all influences except the Sun, is it possible for the grain to move away from the Sun?

    1. Yes, if the grain is sufficiently large and is a good absorber of light.

    2. Yes, if the grain is sufficiently small and is a good absorber of light.

    3. No, the Sun's gravitational field always attracts the grain to the Sun.

GEOMETRICAL OPTICS I

1. Snell's law describes

    1. Huygens' construction.

    2. magnification.

    3. reflection.

    4. refraction.

2. The phenomenon of dispersion occurs when

    1. there is total internal reflection.

    2. the index of refraction depends on the wavelength.

    3. there is a virtual image.

    4. the incident beam is completely reflected.

3. For angles of incidence exceeding a certain value, light traveling from a medium of high refractive index to one of lower index is

    1. totally reflected.

    2. dispersed.

3. totally refracted.

4. completely polarized.

4. Light is incident upon two polarizing filters arranged in tandem. The filters are crossed so that their polarization directions are perpendicular. The transmitted intensity through the second filter

1. is 100%.

2. depends on the frequency of the incident light.

3. depends on the intensity of the incident light.

4. is zero.

GEOMETRICAL OPTICS II

1. Light from an object is reflected by a mirror in such a way that the rays diverge from and pass through the reflection. This is known as

1. a virtual image.

2. a real image.

3. spherical aberration.

4. a focal point.

2. Which of the following is *not* a principal ray of a spherical mirror?

1. a ray that goes through the center of the sphere

2. a ray that approaches the mirror along a line parallel to the axis

3. a ray that goes through the focal point on the way to the mirror

4. a ray that hits the mirror at the same place that the axis hits

3. For a lens that produces a positive magnification, the image is

1. virtual and upright.

2. virtual and inverted.

3. real and upright.

4. real and inverted.

4. For a thin lens made of two spherical surfaces, the focal length given by the lens-maker's formula depends on

1. the index of refraction of the lens.

2. the radii of the two spherical surfaces.

3. the assumption of incident rays near the axial line.

4. the magnification of the lens.
5. all of the above.
6. 1 and 2.
7. 1, 2, and 3.

PHYSICAL OPTICS I

1. Interference occurs with
   1. light waves.
   2. sound waves.
   3. water waves.
   4. all of the above.
   5. none of the above.

2. In order for interference effects to be observable,
   1. the wavelength of the light must be comparable to the width of any apertures the light encounters.
   2. the intensity of the light must be sufficiently high.
   3. the phase relationships between waves is not important.
   4. the wavelength of the light must be much smaller than the width of any apertures the light encounters.

3. If the interference pattern produced by two light sources is to remain stationary in space, the sources must have
   1. different frequencies and an arbitrary phase difference.
   2. the same frequencies and an arbitrary phase difference.
   3. different frequencies and a phase difference that is time-independent.
   4. the same frequencies and a phase difference that is time-independent.

4. Which term does not arise in the discussion of interference patterns?
   1. coherent sources
   2. Fraunhofer approximation
   3. magnifying power
   4. principal maximum

PHYSICAL OPTICS II

1. The bending of light around an obstacle is called
   1. interference.
   2. resolution.
   3. diffraction.
   4. coherence.

2. Light impinges on a single slit but suffers no significant diffraction. We conclude that the wavelength of the light is
   1. much shorter than the slit width.
   2. much longer than the slit width.
   3. on the order of the slit width.
   4. We cannot say anything about the wavelength.

DIFFRACTION

1. The diffraction pattern generated by a single slit can be constructed using the
   1. Fresnel approximation.
   2. Huygens-Fresnel principle.
   3. Huygens construction.
   4. Rayleigh criterion.

2. Light waves from two point-like sources arrive at the circular aperture of a telescope simultaneously. The telescope will resolve the two sources if which of the following conditions is satisfied?
   1. the Fresnel approximation
   2. the Fraunhofer approximation
   3. the Huygens-Fresnel principle
   4. the Rayleigh criterion

HISTORICAL INTRODUCTION TO MODERN PHYSICS

1. The spectral emittance of a blackbody depends on
   1. the material out of which the body is made.
   2. the characteristics of the body's surface.
   3. the body's temperature.
   4. all of the above.

2. Calculated classically, the spectral emittance of a blackbody diverges at short wavelengths. This result is known as
   1. the Stefan-Boltzmann law.
   2. the ultraviolet catastrophe.
   3. the Compton effect.
   4. Wien's law.

3. The number of photoelectrons emitted from a metal surface depends on
   1. the frequency of the incident light.
   2. the workfunction of the metal.
   3. both of the above.

4. neither of the above.

4. As the wavelength of the light incident on a metal surface is shortened, the kinetic energy of photoelectrons emitted from the surface
   1. increases.
   2. decreases.
   3. stays the same.
   4. need more information

WAVE-PARTICLE DUALITY/UNCERTAINTY

1. The Compton effect illustrates
   1. the wave nature of light.
   2. the ejection of an electron from an irradiated metal surface.
   3. the particle nature of light.
   4. the probabilistic nature of quantum waves.

2. In the Compton experiment, the wavelength of the scattered light is
   1. longer than
   2. the same as
   3. shorter than
   the wavelength of the incident light.

3. The probability of finding a photon of light at a given point
   1. increases as the wavelength of the light decreases.
   2. is proportional to the intensity of the light.
   3. is proportional to the magnitude of the electric field.
   4. is independent of the electric field.

4. Suppose the momentum of a photon is determined with complete accuracy (the uncertainty approaches zero). The uncertainty in a simultaneous measurement of the photon's position
   1. also approaches zero.
   2. approaches infinity.
   3. has some intermediate value.
   4. cannot be determined.

SPECTRAL LINES

1. White light passes through sodium vapor and is then analyzed with a prism. The resulting spectrum
    1. is continuous.
    2. consists of spectral lines.
    3. is continuous and contains absorption lines.
    4. none of the above

2. The systematic pattern in the spacing of the spectral lines of hydrogen was fit to an empirical formula by
    1. Balmer.
    2. de Broglie.
    3. Bohr.
    4. Rutherford.

3. The Rutherford alpha particle/gold foil experiment gave evidence for the
    1. existence of matter waves.
    2. Rydberg-Ritz combination principle.
    3. "plum-pudding" model of the atom.
    4. nuclear atom.

BOHR ATOM

1. Which quantity(ies) is(are) quantized in the Bohr atom?
    1. the electron orbit
    2. the electron energy
    3. the electron angular momentum
    4. all of the above
    5. two of the above

2. In the Bohr atom, the laws of classical mechanics apply to
    1. the orbital motion of the electron in a stationary state.
    2. the motion of the electron during transitions between stationary states.
    3. both of the above.
    4. neither of the above.

3. In the Bohr atom, an electron radiates

1. when accelerating in its orbit around the nucleus.

2. during transitions between orbits.

3. both of the above

4. neither of the above

4. Who postulated the wavelike properties of material particles?

1. Bohr

2. Schrodinger

3. Heisenberg

4. de Broglie

ANSWER KEY FOR READING QUIZZES

Note: Topic followed by question numbers and corresponding answers.

1. Kinematics: 1-4, 2-1, 3-2, 4-2
2. Newton's Laws: 1-3, 2-2, 3-1
3. Forces: 1-4, 2-1, 3-2
4. Work: 1-3, 2-3, 3-1, 4-3
5. Conservative Forces: 1-1, 2-2, 3-4
6. Potential Energy: 1-1, 2-2, 3-4
7. Gravitation: 1-3, 2-4, 3-3, 4-4
8. Momentum: 1-4, 2-2, 3-3, 4-2
9. Collisions: 1-2, 2-4, 3-3, 4-2
10. Rotational Kinematics I: 1-2, 2-2, 3-2
11. Rotational Kinematics II: 1-4, 2-2, 3-3
12. Rotational Dynamics I: 1-2, 2-2, 3-3
13. Rotational Dynamics II: 1-3, 2-3, 3-1
14. Oscillations: 1-2, 2-3, 3-3, 4-3
15. Waves: 1-1, 2-2, 3-2, 4-2
16. Sound: 1-5, 2-3, 3-2, 4-3
17. Fluid Statics: 1-2, 2-3, 3-1, 4-2
18. Fluid Dynamics: 1-1, 2-2, 3-4, 4-3
19. Electrostatics I: 1-3, 2-2, 3-2, 4-2
20. Electrostatics II: 1-4, 2-2, 3-2, 4-4
21. Electric Potential I: 1-1, 2-3, 3-3
22. Electric Potential II: 1-3, 2-2, 3-1
23. Capacitance: 1-1, 2-3, 3-1, 4-5
24. Ohm's Law: 1-3, 2-3, 3-2, 4-1
25. DC Circuits: 1-3, 2-5, 3-3, 4-2
26. Magnetostatics: 1-2, 2-5, 3-5, 4-2
27. Ampère's Law: 1-3, 2-5
28. Hall Effect: 1-3, 2-1
29. Magnetic Inductance: 1-1, 3-3
30. Mutual Inductance: 1-5, 2-1
31. AC Circuits I: 1-3, 2-2, 3-1, 4-3
32. AC Circuits II: 1-3, 3-2, 3-4, 4-3
33. Maxwell's Equations: 1-3, 2-4
34. Electromagnetic Waves I: 1-1, 2-1
35. Electromagnetic Waves II: 1-5, 2-2
36. Geometrical Optics I: 1-4, 2-2, 3-1, 4-4
37. Geometrical Optics II: 1-2, 2-4, 3-1, 4-7
38. Physical Optics I: 1-4, 2-1, 3-4, 4-3
39. Physical Optics II: 1-3, 2-1
40. Diffraction: 1-2, 2-4
41. Historical Introduction to Modern Physics: 1-3, 2-2, 3-3, 4-1
42. Wave-Particle Duality/Uncertainty: 1-3, 2-1, 3-2, 4-2
43. Spectral Lines: 1-3, 2-1, 4-4
44. Bohr Atom: 1-4, 2-1, 3-2, 4-4

11

CONCEPTESTS

On the following pages are a full set of *ConcepTests*[1] for a one-year intro-ductory physics course. Each question is followed by a line indicating the concepts covered and a brief explanation. The purpose of these explanations is to outline the verbal arguments I use to explain the correct choice of answer. All of the *ConcepTests* are in a ready-to-print format on the diskette. They can be either printed directly on overhead transparencies or cut and pasted to suit your classroom needs. The *ConcepTests* may not otherwise be distributed or edited without written permission from the publisher.

Note: For questions with *numbered* choices, only *one* single choice required; for questions with *alphabetical* choices, choose all that apply.

[1]Courtesy, Eric Mazur, Michael Aziz, William Paul, and Deborah Alpert, Harvard University. Questions 5 and 12 on Magnetism from *A Guide to Introductory Physics Teaching*, Arnold G. Arons, ©1990. Reprinted by permission of John Wiley & Sons, Inc.

KINEMATICS

1. A person initially at point P in the illustration stays there a moment and then moves along the axis to Q and stays there a moment. She then runs quickly to R, stays there a moment, and then strolls slowly back to P. Which of the position vs. time graphs below correctly represents this motion?

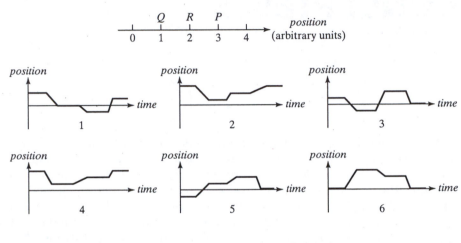

Answer: 2.

2. An object goes from one point in space to another. After it arrives at its destination, its displacement is:
    1. either greater than or equal to
    2. always greater than
    3. always equal to
    4. either smaller than or equal to
    5. always smaller than
    6. either smaller or larger

    than the distance it traveled.

    Answer: 4. The displacement is the vector pointing from the final and to the initial position (and can be negative). The distance traveled is the absolute length of the path traversed; it is always positive and can be larger than the displacement.

3. A marathon runner runs at a steady 15 km/hr. When the runner is 7.5 km from the finish, a bird begins flying from the runner to the finish at 30 km/hr. When the bird reaches the finish line, it turns around and flies back

to the runner, and then turns around again, repeating the back-and-forth trips until the runner reaches the finish line. How many kilometers does the bird travel?

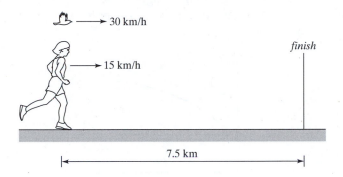

1. 10 km
2. 15 km
3. 20 km
4. 30 km

Answer: 2. It takes the runner half an hour to reach the finish.

4. If you drop an object in the absence of air resistance, it accelerates downward at 9.8 m/s^2. If instead you throw it downward, its downward acceleration after release is

1. less than 9.8 m/s^2.
2. 9.8 m/s^2.
3. more than 9.8 m/s^2.

Answer: 2. The acceleration of gravity is a constant, independent of initial velocity.

5. A person standing at the edge of a cliff throws one ball straight up and another ball straight down at the same initial speed. Neglecting air resistance, the ball to hit the ground below the cliff with the greater speed is the one initially thrown

1. upward.
2. downward.
3. neither—they both hit at the same speed.

Answer: 3. Upon its descent, the velocity of an object thrown straight up with an initial velocity v is exactly $-v$ when it passes the point at which it was first released.

6. A train car moves along a long straight track. The graph shows the position as a function of time for this train. The graph shows that the train:

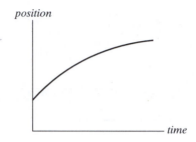

1. speeds up all the time.
2. slows down all the time.
3. speeds up part of the time and slows down part of the time.
4. moves at a constant velocity.

Answer: 2. The slope of the curve diminishes as time increases.

7. The graph shows position as a function of time for two trains running on parallel tracks. Which is true:

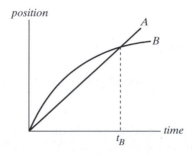

1. At time t_B, both trains have the same velocity.
2. Both trains speed up all the time.
3. Both trains have the same velocity at some time before t_B.
4. Somewhere on the graph, both trains have the same acceleration.

Answer: 3. The slope of curve B is parallel to line A at some point $t < t_B$.

8. You are throwing a ball straight up in the air. At the highest point, the ball's
   1. velocity and acceleration are zero.
   2. velocity is nonzero but its acceleration is zero.
   3. acceleration is nonzero, but its velocity is zero.
   4. velocity and acceleration are both nonzero.

   Answer: 3. The ball reaches its highest point when its velocity is zero; the acceleration of gravity is never zero.

9. A cart on a roller-coaster rolls down the track shown below. As the cart rolls beyond the point shown, what happens to its speed and acceleration in the direction of motion?

   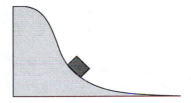

   1. Both decrease.
   2. The speed decreases, but the acceleration increases.
   3. Both remain constant.
   4. The speed increases, but acceleration decreases.
   5. Both increase.
   6. Other

   Answer: 4. Because the height of the track is always decreasing, the cart always accelerates and its speed increases. Because the steepness of the curve decreases with decreasing height, the acceleration decreases.

10. A ball is thrown vertically up, its speed slowing under the influence of gravity. Suppose (*a*) we film this motion and play the tape backward (so the tape begins with the ball at its highest point and ends with it reaching the point from which it was released), and (*b*) we observe the motion of the ball from a frame of reference moving up at the initial speed of the ball. The ball has a downward acceleration g in
    1. (*a*) and (*b*).

2. only (*a*).

3. only (*b*).

4. neither (*a*) nor (*b*).

Answer: 1. Frames (*a*) and (*b*) are both inertial, and thus changes in velocity are the same as in Earth's reference frame.

11. Consider the situation depicted here. A gun is aimed directly at a dangerous criminal hanging from the gutter of a building. The target is well within the gun's range, but the instant the gun is fired and the bullet moves with a speed v_o, the criminal lets go and drops to the ground. What happens? The bullet

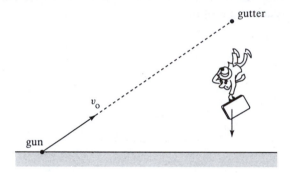

1. hits the criminal regardless of the value of v_o.

2. hits the criminal only if v_o is large enough.

3. misses the criminal.

Answer: 1. The downward acceleration of the bullet and the criminal are identical, so the bullet will hit its target—they both "fall" the same distance.

12. A battleship simultaneously fires two shells at enemy ships. If the shells follow the parabolic trajectories shown, which ship gets hit first?

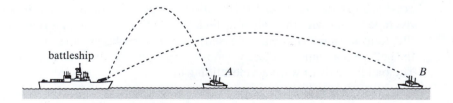

1. *A*
2. both at the same time
3. *B*
4. need more information

Answer: 3. The time a projectile spends in the air is equal to twice the time it takes to fall from its maximum height. Because the shell fired at ship *A* reaches a higher altitude than the one aimed at *B*, the former takes longer to return to sea level.

FORCES

1. A constant force is exerted on a cart that is initially at rest on an air track. Friction between the cart and the track is negligible. The force acts for a short time interval and gives the cart a certain final speed.

To reach the same final speed with a force that is only half as big, the force must be exerted on the cart for a time interval

1. four times as long as
2. twice as long as
3. equal to
4. half as long as
5. a quarter of

that for the stronger force.

Answer: 2. The final speed is proportional to the acceleration of the cart and the time over which it acts.

2. A constant force is exerted for a short time interval on a cart that is initially at rest on an air track. This force gives the cart a certain final speed. The same force is exerted for the same length of time on another cart, also initially at rest, that has twice the mass of the first one. The final speed of the heavier cart is

1. one-fourth

2. four times

3. half

4. double

5. the same as

that of the lighter cart.

Answer: 3. The final speed is proportional to both the force on the cart and the time over which it acts, and inversely proportional to the mass of the cart.

3.   A constant force is exerted for a short time interval on a cart that is initially at rest on an air track. This force gives the cart a certain final speed. Suppose we repeat the experiment but, instead of starting from rest, the cart is already moving with constant speed in the direction of the force at the moment we begin to apply the force. After we exert the same constant force for the same short time interval, the increase in the cart's speed

1. is equal to two times its initial speed.

2. is equal to the square of its initial speed.

3. is equal to four times its initial speed.

4. is the same as when it started from rest.

5. cannot be determined from the information provided.

Answer: 4. The increase in speed is proportional to both the force on the cart and the time over which it acts.

4.   Consider a person standing in an elevator that is accelerating upward. The upward normal force N exerted by the elevator floor on the person is

1. larger than

2. identical to

3. smaller than

the downward weight W of the person.

Answer: 1. In order for the person to be accelerated upward, the normal force exerted by the elevator floor on her must exceed her weight.

5. A person pulls a box across a floor. Which is the correct analysis of the situation?

   1. The box moves forward because the person pulls forward slightly harder on the box than the box pulls backward on the person.

   2. Because action always equals reaction, the person cannot pull the box— the box pulls backward just as hard as the person pulls forward, so there is no motion.

   3. The person gets the box to move by giving it a tug during which the force on the box is momentarily greater than the force exerted by the box on the person.

   4. The person's force on the box is as strong as the force of the box on the person, but the frictional force on the person is forward and large while the backward frictional force on the box is small.

   5. The person can pull the box forward only if it weighs more than the box.

   Answer: 4. The force exerted by the person on the box is equal to that exerted by the box on the person. The person moves forward because of a forward frictional force exerted by the floor. The frictional force exerted by the floor on the box is much smaller.

6. A car rounds a curve while maintaining a constant speed. Is there a net force on the car as it rounds the curve?

   1. No—its speed is constant.

   2. Yes.

   3. It depends on the sharpness of the curve and the speed of the car.

   Answer: 2. Acceleration is a change in the speed and/or direction of an object. Thus, because its direction has changed, the car has accelerated and a force must have been exerted on it.

7. In the 17th century, Otto von Güricke, a physicist in Magdeburg, fitted two hollow bronze hemispheres together and removed the air from the resulting sphere with a pump. Two eight-horse teams could not pull the halves apart even though the hemispheres fell apart when air was re-

halves apart even though the hemispheres fell apart when air was re-admitted. Suppose von Güricke had tied both teams of horses to one side and bolted the other side to a heavy tree trunk. In this case, the tension on the hemispheres would be

1. twice
2. exactly the same as
3. half

what it was before.

Answer: 1. If both teams pull on the hemispheres from one side with twice the force and the hemispheres remain at rest because the other side is bolted to a tree trunk, the tree must exert an equal and opposite force. Thus, the tension is doubled.

8. You are pushing a wooden crate across the floor at constant speed. You decide to turn the crate on end, reducing by half the surface area in contact with the floor. In the new orientation, to push the same crate across the same floor with the same speed, the force that you apply must be about

1. four times as great
2. twice as great
3. equally great
4. half as great
5. one-fourth as great

as the force required before you changed the crate's orientation.

Answer: 3. The force is proportional to the coefficient of kinetic friction and the weight of the crate. Neither depends on the size of the surface in contact with the floor.

9. An object is held in place by friction on an inclined surface. The angle of inclination is increased until the object starts moving. If the surface is kept at this angle, the object

1. slows down.
2. moves at uniform speed.
3. speeds up.
4. none of the above

Answer: 3. As the tilt of the surface is increased at a certain angle the object starts sliding. Until that angle is reached, the object is at rest, and the net force on it is zero. For the object to start sliding from rest, there must be a net force on it; if the net force is no longer zero, the object will accelerate.

10. You are a passenger in a car and not wearing your seat belt. Without increasing or decreasing its speed, the car makes a sharp left turn, and you find yourself colliding with the right-hand door. Which is the correct analysis of the situation?

1. Before and after the collision, there is a rightward force pushing you into the door.

2. Starting at the time of collision, the door exerts a leftward force on you.

3. both of the above

4. neither of the above

Answer: 2. Because of the law of inertia, a body has a tendency to continue traveling in a straight line. As it tries to do so, it collides with the door because the door, being part of the car, is beginning to curve leftward. This contact forces the body to go left into the turn.

11. Consider a horse pulling a buggy. Is the following statement true?

The weight of the horse and the normal force exerted by the ground on the horse constitute an interaction pair that are always equal and opposite according to Newton's third law.

1. yes

2. no

Answer: 2. The normal force is a repulsive contact force between the ground and the horse. The weight is the gravitational force exerted by Earth on the horse. These two forces are of different origin and do not constitute an interaction pair.

12. Consider a car at rest. We can conclude that the downward gravitational pull of Earth on the car and the upward contact force of Earth on it are equal and opposite because

1. the two forces form an interaction pair.

2. the net force on the car is zero.

3. neither of the above

Answer: 2. These two forces cannot be an interaction pair because they act on the same object. Because the car is at rest, however, its momentum is constant (and zero). Because net force equals the time rate of change in momentum, the net force on the car must be zero. This means that the two forces must be equal and opposite.

ENERGY, WORK, AND CONSERVATION OF ENERGY

1. At the bowling alley, the ball-feeder mechanism must exert a force to push
    the bowling balls up a 1.0-m long ramp. The ramp leads the balls to a chute
    0.5 m above the base of the ramp. Approximately how much force must
    be exerted on a 5.0-kg bowling ball?

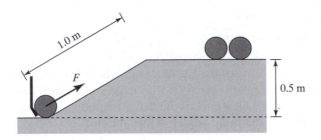

1. 200 N
2. 50 N
3. 25 N
4. 5.0 N
5. impossible to determine

Answer: 3. The force exerted by the mechanism times the distance of 1.0 m
over which the force is exerted must equal the change in the potential energy of the ball.

2. Two marbles, one twice as heavy as the other, are dropped to the ground from
    the roof of a building. Just before hitting the ground, the heavier marble has
    1. as much kinetic energy as the lighter one.
    2. twice as much kinetic energy as the lighter one.
    3. half as much kinetic energy as the lighter one.
    4. four times as much kinetic energy as the lighter one.
    5. impossible to determine

Answer: 2. Kinetic energy is proportional to mass.

3. Suppose you want to ride your mountain bike up a steep hill. Two paths
    lead from the base to the top, one twice as long as the other. Compared
    to the average force exerted along the short path, F_{av}, the average force
    you exert along the longer path is

1. $F_{av}/4$
2. $F_{av}/3$
3. $F_{av}/2$
4. F_{av}
5. undetermined—it depends on the time taken.

Answer: 3. The gravitational potential energy gained is the same in both cases and is equal to the average force exerted times the distance traveled.

4. A piano mover raises a 100-kg piano at a constant rate using the friction-less pulley system shown here. With how much force is he pulling on the rope? Ignore friction and assume $g = 10$ m/s^2.

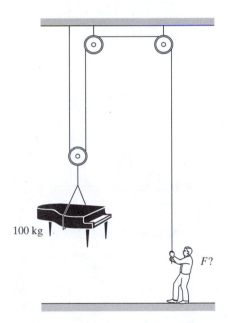

1. 2,000 N
2. 1,500 N
3. 1,000 N
4. 750 N
5. 500 N
6. 200 N
7. 50 N

8. impossible to determine

Answer: 5. The tension in the rope is uniform and, because the piano moves at a constant rate, its weight must equal twice the tension.

5. A 50-kg person stands on a 25-kg platform. He pulls on the rope that is attached to the platform via the frictionless pulley system shown here. If he pulls the platform up at a steady rate, with how much force is he pulling on the rope? Ignore friction and assume $g = 10$ m/s^2.

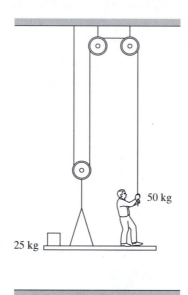

25 kg

50 kg

1. 750 N
2. 625 N
3. 500 N
4. 250 N
5. 75 N
6. 50 N
7. 25 N
8. impossible to determine

Answer: 4. The tension in the rope is uniform. Because of the pulley configuration, the weight of platform plus person must equal three times the tension.

6. A block initially at rest is allowed to slide down a frictionless ramp and attains a speed v at the bottom. To achieve a speed $2v$ at the bottom, how many times as high must a new ramp be?

1. 1
2. 2
3. 3
4. 4
5. 5
6. 6

Answer: 4. The gain in kinetic energy, proportional to the square of the block's speed at the bottom of the ramp, is equal to the loss in potential energy. This, in turn, is proportional to the height of the ramp.

7. A spring-loaded toy dart gun is used to shoot a dart straight up in the air, and the dart reaches a maximum height of 24 m. The same dart is shot straight up a second time from the same gun, but this time the spring is compressed only half as far before firing. How far up does the dart go this time, neglecting friction and assuming an ideal spring?

1. 96 m
2. 48 m
3. 24 m
4. 12 m
5. 6 m
6. 3 m
7. impossible to determine

Answer: 5. The potential energy of a spring is proportional to the square of the distance over which the spring is compressed. All of the spring's potential energy is converted to gravitational potential energy.

8. A sports car accelerates from zero to 30 mph in 1.5 s. How long does it take for it to accelerate from zero to 60 mph, assuming the power of the engine to be independent of velocity and neglecting friction?

1. 2 s
2. 3 s
3. 4.5 s
4. 6 s
5. 9 s
6. 12 s

Answer: 4. In the absence of friction, the power of the engine is equal to the kinetic energy of the car divided by the time it took to attain that kinetic energy.

9. A cart on an air track is moving at 0.5 m/s when the air is suddenly turned off. The cart comes to rest after traveling 1 m. The experiment is repeated, but now the cart is moving at 1 m/s when the air is turned off. How far does the cart travel before coming to rest?

    1. 1 m
    2. 2 m
    3. 3 m
    4. 4 m
    5. 5 m
    6. impossible to determine

    Answer: 4. The cart comes to a stop when all of the cart's kinetic energy is lost to friction. The frictional force times the stopping distance is equal to the cart's initial kinetic energy.

10. Suppose you drop a 1-kg rock from a height of 5 m above the ground. When it hits, how much force does the rock exert on the ground?

    1. 0.2 N
    2. 5 N
    3. 50 N
    4. 100 N
    5. impossible to determine

    Answer: 5. To answer this question, one needs to know how much the ground compresses before the rock comes to a stop.

11. A person pulls a box along the ground at a constant speed. If we consider Earth and the box as our system, what can we say about the net external force on the system?

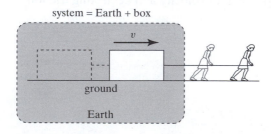

1. It is zero because the system is isolated.
2. It is nonzero because the system is not isolated.
3. It is zero even though the system is not isolated.
4. It is nonzero even though the system is isolated.
5. none of the above

Answer: 3. The Earth plus box system is not isolated because the person pulling the box exerts both a force on the box (she is pulling it) and a force on the Earth (to move forward she must push off the ground). Since Earth is stationary and both the box and the person move at constant speed, the net external force is zero.

12. A person pulls a box along the ground at a constant speed. If we consider Earth and the box as our system, the net force exerted by the person on the system is

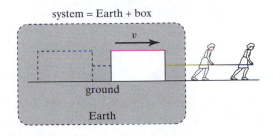

1. zero
2. nonzero

Answer: 1. Below are free-body diagrams for the the box, the person, and Earth. (F_{pb} = force exerted by person on box, etc.)

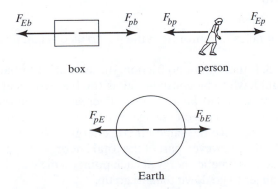

Because of Newton's second and third laws all the forces in the diagram have the same magnitude: $F_{Eb} = F_{pb} = F_{bp} = F_{Ep} = F_{bE} = F_{pE}$. Since the force F_{pE} exerted by the person on Earth is in the opposite direction of the force F_{pb} exerted by the person on the box, the net force on the system (box + Earth) is zero.

13.   A person pulls a box along the ground at a constant speed. If we consider Earth and the box as our system, the work done by the person on the system is:

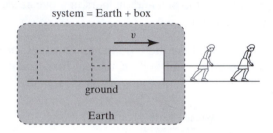

1. zero

2. nonzero

Answer: 2. The displacement of Earth is zero, and so the work done by the person on Earth is zero. The displacement of the box is nonzero, however, and therefore the person does work on the box. The sum of the work done on the box and that done on Earth is then nonzero.

14.   A stone is launched upward into the air. In addition to the force of gravity, the stone is subject to a frictional force due to air resistance. The time the stone takes to reach the top of its flight path is

1. larger than

2. equal to

3. smaller than

the time it takes to return from the top to its original position.

Answer: 3. If there were no friction, the sum of potential and kinetic energy would have to be constant. So as the ball rose, its potential energy would increase and its kinetic energy decrease. On the way down, the potential energy would decrease again such that, at each point on the path, the ball would have the same kinetic energy as on the way up. Because there is friction, however, part of the total energy is dissipated. As a result, the ball has less kinetic energy at each point on the way down. This means it takes longer to go down than to go up.

GRAVITATION

1. Which of the following depends on the inertial mass of an object (as opposed to its gravitational mass)?
    1. the time the object takes to fall from a certain height
    2. the weight of the object on a bathroom-type spring scale
    3. the acceleration given to the object by a compressed spring
    4. the weight of the object on an ordinary balance

    Answer: 3. The other experiments measure the object's gravitational mass.

2. Two satellites A and B of the same mass are going around Earth in concentric orbits. The distance of satellite B from Earth's center is twice that of satellite A. What is the ratio of the centripetal force acting on B to that acting on A?
    1. 1/8
    2. 1/4
    3. 1/2
    4. $\sqrt{1/2}$
    5. 1

    Answer: 2. The centripetal force on each satellite is provided by the gravitational force between the satellite and Earth.

3. Two satellites A and B of the same mass are going around Earth in concentric orbits. The distance of satellite B from Earth's center is twice that of satellite A. What is the ratio of the tangential speed of B to that of A?
    1. $1/2$
    2. $\sqrt{1/2}$
    3. 1
    4. $\sqrt{2}$
    5. 2

    Answer: 2. For each satellite, the centripetal force is equal to the gravitational force between the satellite and Earth, and is proportional to the square of the tangential velocity and the inverse of the distance.

4. Suppose Earth had no atmosphere and a ball were fired from the top of Mt. Everest in a direction tangent to the ground. If the initial speed were high enough to cause the ball to travel in a circular trajectory around Earth, the ball's acceleration would

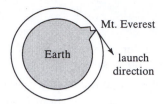

1. be much less than g (because the ball doesn't fall to the ground).
2. be approximately g.
3. depend on the ball's speed.

Answer: 2. The ball's acceleration is caused by the only force exerted on the ball: gravitation. Near the surface of Earth, the value of this acceleration is g (perhaps a little less because of the altitude of Mt. Everest).

5. A rock, initially at rest with respect to Earth and located an infinite distance away is released and accelerates toward Earth. An observation tower is built 3 Earth-radii high to observe the rock as it plummets to Earth. Neglecting friction, the rock's speed when it hits the ground is

   1. twice
   2. three times
   3. four times
   4. six times
   5. eight times
   6. nine times
   7. sixteen times

   its speed at the top of the tower.

   Answer: 1. The increase in the rock's kinetic energy at any point is equal in magnitude to the decrease in the gravitational potential energy at that point. The top of the tower is 4 Earth radii from its center.

6. The Moon does not fall to Earth because

   1. It is in Earth's gravitational field.
   2. The net force on it is zero.
   3. It is beyond the main pull of Earth's gravity.
   4. It is being pulled by the Sun and planets as well as by Earth.
   5. all of the above
   6. none of the above

Answer: 6. The Moon is accelerating toward Earth because of the gravitational attraction between the two. This attraction supplies the centripetal force necessary to keep the Moon in orbit.

7. A pendulum bob is suspended from a long pole somewhere on the northern hemisphere. When the pendulum is at rest, the combined action of gravitation and Earth's rotation makes the bob

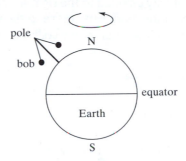

1. point straight down toward the center of Earth.
2. deviate toward the east.
3. deviate toward the west.
4. deviate toward the north.
5. deviate toward the south.
6. none of the above

Answer: 5. If Earth were not rotating, the bob would point straight toward the center of Earth. The rotation of the Earth requires the bob to deviate southward to increase the tension in the string so that the bob has the correct centripetal acceleration.

INERTIAL MASS, MOMENTUM, COLLISIONS

1. An astronaut floating weightlessly in orbit shakes a large iron anvil rapidly back and forth. She reports back to Earth that

    1. the shaking costs her no effort because the anvil has no inertial mass in space.
    2. the shaking costs her some effort but considerably less than on Earth.
    3. although weightless, the inertial mass of the anvil is the same as on Earth.

Answer: 3. Inertial mass is an intrinsic property of any object: It is a quantitative measure of that object's resistance to being accelerated. Weight, on the other hand, is a measure of Earth's gravitational pull on an object. In orbit, objects are weightless because they are in free fall, but they still resist being accelerated exactly as they do on Earth.

2. You are given two carts, *A* and *B*. They look identical, and you are told that they are made of the same material. You place *A* at rest on an air track and give *B* a constant velocity directed to the right so that it collides elastically with *A*. After the collision, both carts move to the right, the velocity of *B* being smaller than what it was before the collision. What do you conclude?

   1. Cart *A* is hollow.
   2. The two carts are identical.
   3. Cart *B* is hollow.
   4. need more information

Answer: 1. Because *B* continues to move in the original direction of motion albeit at a slower speed, its inertial mass must be larger than that of *A*. Since they are made of the same material and of identical shape, *A* must be hollow.

3. Which of these systems are isolated?

   A. While slipping on a patch of ice, a car collides totally inelastically with another car. *System:* both cars
   B. Same situation as in A. *System:* slipping car
   C. A single car slips on a patch of ice. *System:* car
   D. A car makes an emergency stop on a road. *System:* car
   E. A ball drops to Earth. *System:* ball
   F. A billiard ball collides elastically with another billiard ball on a pool table. *System:* both balls

Answer: A, C, and F. Explanation:

   A. When the ice is approximated as being frictionless, the two cars are isolated from external influences.
   B. The slipping car interacts with the other car.
   C. As in A, the single car is isolated from external influences.
   D. The interaction between the car's tires and the road slows the car.
   E. The interaction with Earth accelerates the ball downward.
   F. When friction with the pool table is ignored, the two balls are isolated from external influences.

Note that, strictly speaking, none of the above is truly isolated; we must ignore certain (small) interactions.

4. A car accelerates from rest. In doing so the absolute value of the car's momentum changes by a certain amount and that of the Earth changes by
   1. a larger amount.
   2. the same amount.
   3. a smaller amount.
   4. The answer depends on the interaction between the two.

   Answer: 2. Momentum is equal to the force times the time over which it acts. The forces exerted on the car by Earth and those exerted on Earth by the car are equal and opposite, and the times during which these forces are exerted are equal.

5. A car accelerates from rest. It gains a certain amount of kinetic energy and Earth
   1. gains more kinetic energy.
   2. gains the same amount of kinetic energy.
   3. gains less kinetic energy.
   4. loses kinetic energy as the car gains it.

   Answer: 3. From the work-energy theorem, the kinetic energy gained equals the force times the distance over which it acts. The forces exerted on the car by Earth and those exerted on Earth by the car are equal and opposite, but the distances over which these forces act are not equal—Earth moves a negligible amount as a result of the car's motion.

6. Suppose the entire population of the world gathers in one spot and, at the sounding of a prearranged signal, everyone jumps up. While all the people are in the air, does Earth gain momentum in the opposite direction?
   1. No; the inertial mass of Earth is so large that the planet's change in motion is imperceptible.
   2. Yes; because of its much larger inertial mass, however, the change in momentum of Earth is much less than that of all the jumping people.
   3. Yes; Earth recoils, like a rifle firing a bullet, with a change in momentum equal to and opposite that of the people.
   4. It depends.

   Answer: 3. If we consider Earth to be an isolated system (during the short time interval that the people jump up, this approximation is appropriate), then momentum must be conserved. So the momentum of Earth must be equal to and opposite that of the jumping people. Because of Earth's large inertial mass, however, there is no perceptible motion.

7. Suppose the entire population of the world gathers in one spot and, at the sound of a prearranged signal, everyone jumps up. About a second later, 5 billion people land back on the ground. After the people have landed, Earth's momentum is

   1. the same as what it was before the people jumped.
   2. different from what it was before the people jumped.

   Answer: 1. It's impossible to change the momentum of an isolated system from inside the system.

8. Suppose rain falls vertically into an open cart rolling along a straight horizontal track with negligible friction. As a result of the accumulating water, the speed of the cart

   1. increases.
   2. does not change.
   3. decreases.

   Answer: 3. The water, because it falls vertically, does not change the cart's horizontal momentum. The mass of the cart increases, however, and so its speed decreases.

9. Suppose rain falls vertically into an open cart rolling along a straight horizontal track with negligible friction. As a result of the accumulating water, the kinetic energy of the cart

   1. increases.
   2. does not change.
   3. decreases.

   Answer: 3. The cart's momentum is unchanged but its speed decreases so its kinetic energy decreases as well.

10. Consider these situations:
    (*i*) a ball moving at speed v is brought to rest;
    (*ii*) the same ball is projected from rest so that it moves at speed v;
    (*iii*) the same ball moving at speed v is brought to rest and then projected backward to its original speed.

    In which case(s) does the ball undergo the largest change in momentum?
    1. (*i*)
    2. (*i*) and (*ii*)
    3. (*i*), (*ii*), and (*iii*)

4. (*ii*)

5. (*ii*) and (*iii*)

6. (*iii*)

Answer: 6. Let's say the ball has inertial mass m and velocity v. The decrease in momentum in case (*i*) is $0 - mv = -mv$ (final momentum minus initial momentum). In case (*ii*), we find $mv - 0 = +mv$. In case (*iii*), we have $m(-v) - mv = -2mv$ because the ball's velocity is now in the opposite direction. So the magnitude of the change is greatest in the third case.

11. Consider two carts, of masses m and $2m$, at rest on an air track. If you push first one cart for 3 s and then the other for the same length of time, exerting equal force on each, the momentum of the light cart is

1. four times

2. twice

3. equal to

4. one-half

5. one-quarter

the momentum of the heavy cart.

Answer: 3. Momentum is equal to force times time. Because the forces on the carts are equal, as are the times over which the forces act, the final momenta of the two carts are equal.

12. Consider two carts, of masses m and $2m$, at rest on an air track. If you push first one cart for 3 s and then the other for the same length of time, exerting equal force on each, the kinetic energy of the light cart is

1. larger than

2. equal to

3. smaller than

the kinetic energy of the heavy car.

Answer: 1. Because the momenta of the two carts are equal, the velocity of the light cart must be twice that of the heavy cart. Thus, the kinetic energy of the light cart is twice the kinetic energy of the heavy one.

13. Suppose a ping-pong ball and a bowling ball are rolling toward you. Both have the same momentum, and you exert the same force to stop each. How do the time intervals to stop them compare?

1. It takes less time to stop the ping-pong ball.

2. Both take the same time.

3. It takes more time to stop the ping-pong ball.

Answer: 2. Because force equals the time rate of change of momentum, the two balls lose momentum at the same rate. If both balls initially have the same momentum, it takes the same amount of time to stop them.

14.   Suppose a ping-pong ball and a bowling ball are rolling toward you. Both have the same momentum, and you exert the same force to stop each. How do the distances needed to stop them compare?

1. It takes a shorter distance to stop the ping-pong ball.

2. Both take the same distance.

3. It takes a longer distance to stop the ping-pong ball.

Answer: 3. Because the momenta of the two balls are equal, the ball with the larger velocity has the larger kinetic energy. Being that the ping-pong ball has a smaller inertial mass, it must therefore have the larger kinetic energy. This means more work must be done on the ping-pong ball than on the bowling ball. Because work is the product of force and displacement, the distance to stop the ping-pong ball is greater.

15.   If ball 1 in the arrangement shown here is pulled back and then let go, ball 5 bounces forward. If balls 1 and 2 are pulled back and released, balls 4 and 5 bounce forward, and so on. The number of balls bouncing on each side is equal because

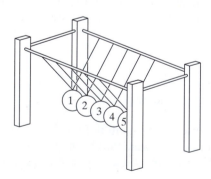

1. of conservation of momentum.

2. the collisions are all elastic.

3. neither of the above

Answer: 2. There are many different final states that conserve momentum, but this is the only one that also conserves kinetic energy. Since conservation of kinetic energy is the same thing as saying "the collisions are elastic," choice 2 is correct.

16.   A cart moving at speed v collides with an identical stationary cart on an airtrack, and the two stick together after the collision. What is their velocity after colliding?

1. v
2. $0.5\,v$
3. zero
4. $-0.5\,v$
5. $-v$
6. need more information

Answer: 2. Because total momentum must be conserved, we have $mv = 2mu$, which gives a final velocity $u = 0.5v$.

17.   A person attempts to knock down a large wooden bowling pin by throwing a ball at it. The person has two balls of equal size and mass, one made of rubber and the other of putty. The rubber ball bounces back, while the ball of putty sticks to the pin. Which ball is most likely to topple the bowling pin?

1. the rubber ball
2. the ball of putty
3. makes no difference
4. need more information

Answer: 1. Because momentum is conserved in these interactions, more momentum is transferred to the bowling pin from the rubber ball than from the putty ball. Hence, the rubber ball is more likely to knock the pin over.

18.   Think fast! You've just driven around a curve in a narrow, one-way street at 25 mph when you notice a car identical to yours coming straight toward you at 25 mph. You have only two options: hitting the other car head on or swerving into a massive concrete wall, also head on. In the split second before the impact, you decide to

1. hit the other car.
2. hit the wall.
3. hit either one—it makes no difference.
4. consult your lecture notes.

Answer: 3. Your change in momentum is the same in both cases. Imagine holding a thin sheet of metal between you and the oncoming car. The sheet will stay put (just as the wall does) because your momentum and that of the other car add up to zero.

19. If all three collisions in the figure shown here are totally inelastic, which bring(s) the car on the left to a halt?

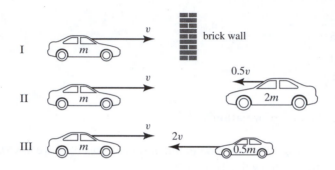

1. I
2. II
3. III
4. I, II
5. I, III
6. II, III
7. all three

Answer: 7. Momentum conservation tells us that all three collisions bring the left-hand car to a halt.

20. If all three collisions in the figure shown are totally inelastic, which cause(s) the most damage?

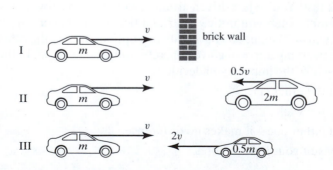

1. I
2. II
3. III
4. I, II
5. I, III
6. II, III
7. all three

Answer: 3. The right car in III loses more kinetic energy in the collision than the right car in II or the wall (which has zero kinetic energy) in I. Since any amount of kinetic energy lost goes into deforming the cars, the most damage occurs in III.

21. A golf ball is fired at a bowling ball initially at rest and bounces back elastically. Compared to the bowling ball, the golf ball after the collision has

1. more momentum but less kinetic energy.
2. more momentum and more kinetic energy.
3. less momentum and less kinetic energy.
4. less momentum but more kinetic energy.
5. none of the above

Answer: 4. The golf ball bounces back at nearly its incident speed, whereas the bowling ball hardly budges. Thus the change in momentum of the golf ball is nearly $-2mv$, and the bowling ball must gain momentun $+2mv$ to conserve momentum. However, since the mass of the bowling ball is much larger than that of the golf ball, the bowling ball's velocity and hence its kinetic energy are much smaller than those of the golf ball.

22. A golf ball is fired at a bowling ball initially at rest and sticks to it. Compared to the bowling ball, the golf ball after the collision has

1. more momentum but less kinetic energy.
2. more momentum and more kinetic energy.
3. less momentum and less kinetic energy.
4. less momentum but more kinetic energy.
5. none of the above

Answer: 3. Both balls move at the same speed after the collision. So the ball with the larger inertial mass has both the larger momentum and the larger kinetic energy.

23.   Suppose you are on a cart, initially at rest on a track with very little friction. You throw balls at a partition that is rigidly mounted on the cart. If the balls bounce straight back as shown in the figure, is the cart put in motion?

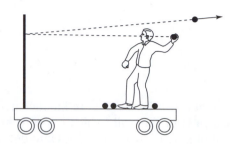

1. Yes, it moves to the right.
2. Yes, it moves to the left.
3. No, it remains in place.

Answer: 2. Because all the balls bounce back to the right, then in order to conserve momentum, the cart must move forward.

24.   A compact car and a large truck collide head on and stick together. Which undergoes the larger momentum change?
1. car
2. truck
3. The momentum change is the same for both vehicles.
4. Can't tell without knowing the final velocity of combined mass.

Answer: 3. Conservation of momentum tells us that the changes in momentum must add up to zero. So the change in the car's momentum must be equal to the change in the truck's momentum, and the two changes must be in the opposite directions.

25.   A compact car and a large truck collide head on and stick together. Which vehicle undergoes the larger acceleration during the collision?
1. car
2. truck
3. Both experience the same acceleration.
4. Can't tell without knowing the final velocity of combined mass.

Answer: 1. Say the car has inertial mass m and the truck has inertial mass $M \gg m$. Because the changes in momentum are equal (neglecting the fact that they are in opposite directions), we have $m\Delta v = M\Delta V$, where Δv

is the change in the car's speed and ΔV the change in the truck's speed. Because $m << M$, $\Delta v >> \Delta V$. The acceleration is proportional to the change in speed, and both changes in speed take place over the same time interval (the duration of the collision). Therefore the car undergoes a much larger acceleration than the truck.

26. Is it possible for a stationary object that is struck by a moving object to have a larger final momentum than the initial momentum of the incoming object?

    1. Yes.
    2. No because such an occurrence would violate the law of conservation of momentum.

    Answer: 1. Think of a small ball bouncing back upon striking a much more massive object. If the ball bounces back at its incoming speed, its change in momentum is $2mv$. Conservation of momentum then requires the more massive object to have a momentum $2mv$ in the opposite direction. This is larger than the initial momentum of the ball, which was mv.

INTERACTIONS

1. Two carts of identical inertial mass are put back-to-back on a track. Cart A has a spring-loaded piston; cart B is entirely passive. When the piston is released, it pushes against cart B, and

   1. A is put in motion but B remains at rest.
   2. both carts are set into motion, with A gaining more speed than B.
   3. both carts gain equal speed but in opposite directions.
   4. both carts are set into motion, with B gaining more speed than A.
   5. B is put in motion but A remains at rest.

   Answer: 3. This is the only answer that satisfies conservation of momentum.

2. Two carts are put back-to-back on a track. Cart A has a spring-loaded piston; cart B, which has twice the inertial mass of cart A, is entirely passive. When the piston is released, it pushes against cart B, and the carts move apart. How do the magnitudes of the final momenta and kinetic energies compare?

   1. $p_A > p_B, k_A > k_B$
   2. $p_A > p_B, k_A = k_B$
   3. $p_A > p_B, k_A < k_B$
   4. $p_A = p_B, k_A > k_B$
   5. $p_A = p_B, k_A = k_B$

6. $p_A = p_B, k_A < k_B$

7. $p_A < p_B, k_A > k_B$

8. $p_A < p_B, k_A = k_B$

9. $p_A < p_B, k_A < k_B$

Answer: 4. Conservation of momentum requires the momenta of the two carts to be equal in magnitude (but in opposite directions). This restriction eliminates all but choices 4–6. Because the inertial mass of B is twice that of A, the velocity of B must be half that of A. Because kinetic energy is momentum times one-half the velocity and because the momenta of the two carts are equal, the cart having the higher speed (A) has the larger kinetic energy.

3.   Two carts are put back-to-back on a track. Cart A has a spring-loaded piston; cart B, which has twice the inertial mass of cart A, is entirely passive. When the piston is released, it pushes against cart B, and the carts move apart. Ignoring signs, while the piston is pushing,

1. A has a larger acceleration than B.

2. the two have the same acceleration.

3. B has a larger acceleration than A.

Answer: 1. Cart A gains more speed than cart B. Because both carts gain speed over the same amount of time, the acceleration of A must be larger than that of B.

4.   Two people on roller blades throw a ball back and forth. After a couple of throws, they are (ignore friction)

1. standing where they were initially.

2. standing farther away from each other.

3. standing closer together.

4. moving away from each other.

5. moving toward each other.

Answer: 4. Each throw is an explosive collision. Let's assume the two people, A and B, start at rest. As A throws the ball, he recoils and moves backward—away from B—so that the total momentum of (ball + A) is conserved. When B catches the ball, she too moves backward, away from A. When she throws the ball to A, she recoils and moves backward even faster. A then catches the ball and receives some more momentum from it. Upon throwing the ball, A gains even more, and on and on. The net effect is that A and B accelerate away from each other.

5. Two people on roller blades throw a ball back and forth. Which statement(s) is/are true?

   A. The interaction mediated by the ball is repulsive.
   B. If we film the action and play the movie backward, the interaction appears attractive.
   C. The total momentum of the two people is conserved.
   D. The total energy of the two people is conserved.

   Answer: A. Because it causes the two interacting bodies to accelerate away from each other, this interaction is repulsive. Statement B is false because if a movie of this event played backward, we would see *A* and *B* moving toward each other. As they get closer, however, they slow down more and more, which means their accelerations are still pointing away from each other. So the interaction is still repulsive.

   What about the conservation laws? The ball carries both momentum and energy back and forth between the two roller-bladers. Their momentum and energy therefore cannot be conserved.

6. In the following figure, a 10-kg weight is suspended from the ceiling by a spring. The weight-spring system is at equilibrium with the bottom of the weight about 1 m above the floor. The spring is then stretched until the weight is just above the eggs. When the spring is released, the weight is pulled up by the contracting spring and then falls back down under the influence of gravity. On the way down, it

   1. reverses its direction of travel well above the eggs.

    2. reverses its direction of travel precisely as it reaches the eggs.

    3. makes a mess as it crashes into the eggs.

Answer: 2. In the initial situation, with the spring stretched until the bottom of the weight is just above the eggs, all of the energy of the system is potential. Upon release, some of this energy is converted to kinetic energy in the moving spring and weight. When the weight returns to its position just above the eggs, however, all of the energy in the system must again be potential. Having no kinetic energy, the weight stops at the point from which it was initially released. The eggs are safe.

7.   In part (*a*) of the figure, an air track cart attached to a spring rests on the track at the position $x_{equilibrium}$ and the spring is relaxed. In (*b*), the cart is pulled to the position x_{start} and released. It then oscillates about $x_{equilibrium}$. Which graph correctly represents the potential energy of the spring as a function of the position of the cart?

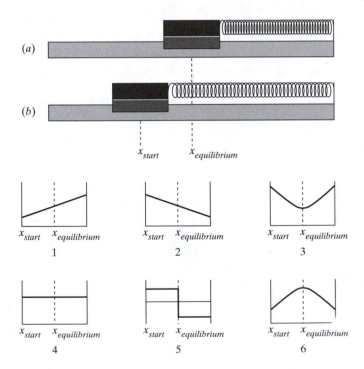

Answer: 3. The cart starts at x_{start} with no kinetic energy, and so the spring's potential energy is a maximum. Once released, the cart accelerates to the right and its kinetic energy increases as the potential energy of the spring is converted into kinetic energy of the cart. As the cart passes the equilibrium position, its kinetic energy is a maximum and so the spring's potential energy is a minimum. Once to the right of $x_{equilibrium}$, the cart starts to compress the spring and it slows down as its kinetic energy is converted back to potential energy of the recompressed spring. At the rightmost point it reaches, the cart reverses its direction of travel. At that instant, it has no kinetic energy and the spring again has maximum potential energy.

8. Two cars, one twice as heavy as the other, are at rest on a horizontal track. A person pushes each car for 5 s. Ignoring friction and assuming equal force exerted on both cars, the momentum of the light car after the push is

    1. smaller than

    2. equal to

    3. larger than

    the momentum of the heavy car.

    Answer: 2. The change in momentum caused by a constant force is the product of the force and the time interval: $\Delta p = F \Delta t$. Because the time interval Δt and the force F are the same for both cars, the changes in momentum are also equal.

9. Two cars, one twice as heavy as the other, are at rest on a horizontal track. A person pushes each car for 5 s. Ignoring friction and assuming equal force exerted on both cars, the kinetic energy of the light car after the push is

    1. smaller than

    2. equal to

    3. larger than

    the kinetic energy of the heavy car.

    Answer: 3. Because the momenta of the two cars are equal, the car with the larger velocity must have the larger kinetic energy. This will be the lighter of the two; because it has less inertia, its acceleration is larger that that of the heavy car.

REFERENCE FRAMES

1. When a small ball collides elastically with a more massive ball initially at rest, the massive ball tends to remain at rest, whereas the small ball bounces back at almost its original speed. Now consider a massive ball of inertial mass M moving at speed v and striking a small ball of inertial mass m initially at rest. The change in the small ball's momentum is

1. Mv
2. $2Mv$
3. mv
4. $2mv$
5. none of the above

Answer: 4. Consider the second collision from a reference frame moving along with the massive ball. In this moving frame, the massive ball is at rest and the small ball moves toward the massive one at velocity $-v$. This, however, is precisely the first collision. After the collision, the small ball bounces back at velocity v. In the frame of Earth, this corresponds to a speed of $v + v = 2v$. So the change in momentum is $2mv$, exactly as in the first collision.

2. A small rubber ball is put on top of a volleyball, and the combination is dropped from a certain height. Compared to the speed it has just before the volleyball hits the ground, the speed with which the rubber ball rebounds is

   1. the same.
   2. twice as large.
   3. three times as large.
   4. four times as large.
   5. none of the above

   Answer: 3. When the volleyball hits the ground, it reverses its velocity and v becomes $-v$. In a reference frame moving upward along with the volleyball at velocity $-v$, the volleyball is at rest and the rubber ball comes in at speed $2v$. When it hits the volleyball, the rubber ball reverses its speed, and so $2v$ becomes $-2v$. This value for the rubber ball's velocity is valid in the frame moving at velocity $-v$; in Earth's frame, the velocity of the small ball is $-2v + (-v) = -3mv$.

3. Suppose you are sitting in a soundproof, windowless room aboard a hovercraft moving over flat terrain. Which of the following can you detect from inside the room?

   1. rotation
   2. deviation from the horizontal orientation
   3. motion at a steady speed
   4. acceleration
   5. state of rest with respect to ground

Answer: 1, 2, and 4. There are no experiments that can detect uniform motion (or rest); we can sense any motion that causes acceleration.

4. An air track cart initially at rest is put in motion when a compressed spring is released and pushes the cart. In the frame of reference of Earth, the velocity-versus-time graph of the cart is shown here. In a frame moving at constant speed relative to Earth, the cart's change in the following quantities can have any value:

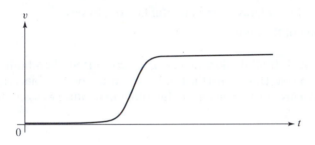

1. velocity
2. momentum
3. kinetic energy
4. none of the above

Answer: 3. Changes in velocity and momentum are the same in any two inertial frames moving at constant speed relative to one another. The change in kinetic energy, however, can take on any value.

5. An air track cart initially at rest is put in motion when a compressed spring is released and pushes the cart. Earth and the cart constitute an isolated system. The change in the cart's kinetic energy is different in the frame of reference of Earth and in a frame moving at constant speed relative to Earth because in the moving frame:

1. conservation of energy does not apply.
2. the amount of energy released by the spring is different.
3. the change in the kinetic energy of Earth is different.
4. a combination of 2 and 3.
5. none of the above

Answer: 3. Because the system is isolated, conservation of momentum holds in any inertial frame. The amount of energy released by the spring is a measure for the change in its physical state, which must be independent of the reference frame. Option 3 is the only option that satisfies conservation of energy.

6. Two objects collide inelastically. Can all the initial kinetic energy in the collision be converted to other forms of energy?
    1. Yes, but only for certain special initial speeds.
    2. Yes, provided the objects are soft enough.
    3. No, this violates a fundamental law of physics.
    4. none of the above

Answer: 1. If all the kinetic energy is converted, then both objects must come to rest. This means the total momentum of the objects after collision is zero. This can happen only if the total momentum was zero to begin with.

ROTATIONS

1. A ladybug sits at the outer edge of a merry-go-round, and a gentleman bug sits halfway between her and the axis of rotation. The merry-go-round makes a complete revolution once each second. The gentleman bug's angular speed is

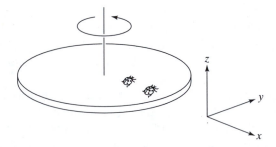

    1. half the ladybug's.
    2. the same as the ladybug's.
    3. twice the ladybug's.
    4. impossible to determine

Answer: 2. Both insects have an angular speed of 1 rev/s.

2. A ladybug sits at the outer edge of a merry-go-round, that is turning and slowing down. At the instant shown in the figure, the radial component of the ladybug's (Cartesian) acceleration is

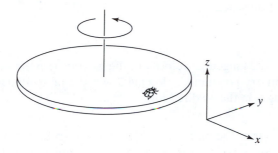

1. in the $+x$ direction.
2. in the $-x$ direction.
3. in the $+y$ direction.
4. in the $-y$ direction.
5. in the $+z$ direction.
6. in the $-z$ direction.
7. zero.

Answer: 2. The radial component of the Cartesian acceleration of a rotating object points toward the axis of rotation.

3. A ladybug sits at the outer edge of a merry-go-round that is turning and slowing down. The tangential component of the ladybug's (Cartesian) acceleration is

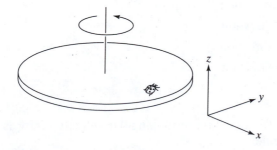

1. in the $+x$ direction.
2. in the $-x$ direction.

3. in the +y direction.

4. in the −y direction.

5. in the +z direction.

6. in the −z direction.

7. zero.

Answer: 4. The tangential velocity of the ladybug is in the +y direction. Because the merry-go-round is slowing down, the tangential acceleration of the ladybug is in the −y direction.

4. A ladybug sits at the outer edge of a merry-go-round that is turning and is slowing down. The vector expressing her angular velocity is

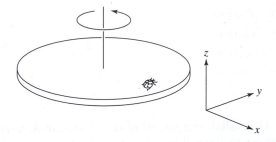

1. in the +x direction.

2. in the −x direction.

3. in the +y direction.

4. in the −y direction.

5. in the +z direction.

6. in the −z direction.

7. zero.

Answer: 5. The direction of the angular velocity vector is given by the right-hand rule.

5. A rider in a "barrel of fun" finds herself stuck with her back to the wall. Which diagram correctly shows the forces acting on her?

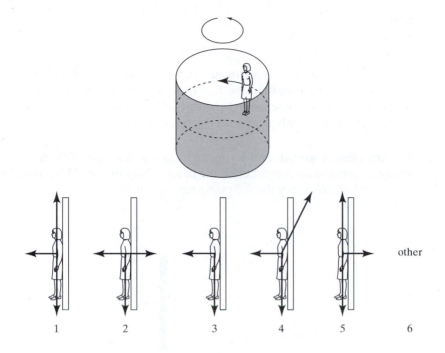

Answer: 1. The normal force of the wall on the rider provides the centripetal acceleration necessary to keep her going around in a circle. The downward force of gravity is equal and opposite to the upward frictional force on her.

6. Consider two people on opposite sides of a rotating merry-go-round. One of them throws a ball toward the other. In which frame of reference is the path of the ball straight when viewed from above: (*a*) the frame of the merry-go-round or (*b*) that of Earth?

1. (*a*) only

2. (*a*) and (*b*)—although the paths appear to curve

3. (*b*) only

4. neither; because it's thrown while in circular motion, the ball travels along a curved path.

Answer: 3. If the reference frame is inertial, the path of the ball must be straight. Of the two frames, only (*b*) is inertial, and so only in (*b*) is the path straight.

7. You are trying to open a door that is stuck by pulling on the doorknob in a direction perpendicular to the door. If you instead tie a rope to the doorknob and then pull with the same force, is the torque you exert increased?

1. yes

2. no

Answer: 2. Because the force you are applying is unchanged and the perpendicular distance between the line of action and the pivot point (the lever arm) is likewise unchanged, the torque you apply does not increase. (Pulling at an angle only decreases the lever arm.)

8. You are using a wrench and trying to loosen a rusty nut. Which of the arrangements shown is most effective in loosening the nut? List in order of descending efficiency the following arrangements:

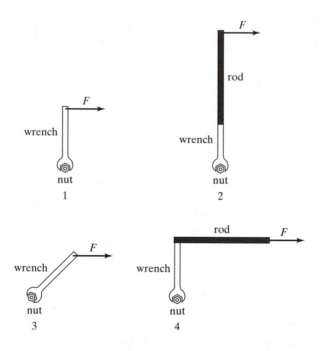

Answer: 2-1-4-3. To increase a torque, you can increase either the applied force or the moment arm. Here the force is the same in all four situations, and so this question boils down to comparing moment arms.

9. A force F is applied to a dumbbell for a time interval Δt, first as in (*a*) and then as in (*b*). In which case does the dumbbell acquire the greater center-of-mass speed?

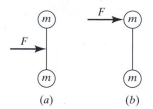

(a)   (b)

1. (a)
2. (b)
3. no difference
4. The answer depends on the rotational inertia of the dumbbell.

Answer: 3. Because the force acts for the same time interval in both cases, the change in momentum ΔP must be the same in both cases, and thus the center-of-mass velocity must be the same in both cases.

10. A force F is applied to a dumbbell for a time interval Δt, first as in (a) and then as in (b). In which case does the dumbbell acquire the greater energy?

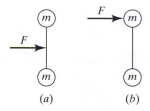

(a)   (b)

1. (a)
2. (b)
3. no difference
4. The answer depends on the rotational inertia of the dumbbell.

Answer: 2. If the center-of-mass velocities are the same, so the translational kinetic energies must be the same. Because dumbbell (b) is also rotating, it also has rotational kinetic energy.

11. Imagine hitting a dumbbell with an object coming in at speed v, first at the center, then at one end. Is the center-of-mass speed of the dumbbell the same in both cases?

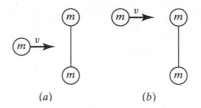

(a)   (b)

1. yes
2. no

Answer: 2. The moving object comes in with a certain momentum. If it hits the center, as in case (*a*), there is no rotation, and this collision is just like a one-dimensional collision between one object of inertial mass *m* and another of inertial mass 2*m*. Because of the larger inertial mass of the dumbbell the incoming object bounces back.

In (*b*), the dumbbell, starts rotating. The incoming object encounters less "resistance" and therefore transfers less of its momentum to the dumbbell, which will therefore have a smaller center-of-mass speed than in case (*a*).

12. A 1-kg rock is suspended by a massless string from one end of a 1-m measuring stick. What is the mass of the measuring stick if it is balanced by a support force at the 0.25-m mark?

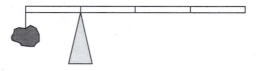

1. 0.25 kg
2. 0.5 kg
3. 1 kg
4. 2 kg
5. 4 kg
6. impossible to determine

Answer: 3. Because the stick is a uniform, symmetric body, we can consider all its weight as being concentrated at the center of mass at the 0.5-m mark. Therefore the point of support lies midway between the two masses, and the system is balanced only if the total mass on the right is also 1 kg.

13. A box, with its center-of-mass off-center as indicated by the dot, is placed on an inclined plane. In which of the four orientations shown, if any, does the box tip over?

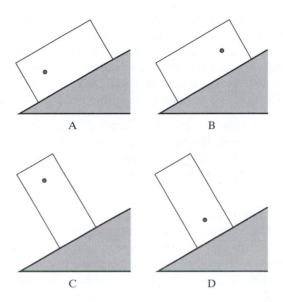

A   B

C   D

Answer: C only. In order to tip over, the box must pivot about its bottom left corner. Only in case C does the torque about this pivot (due to gravity) rotate the box in such a way that it tips over.

14. Consider the situation shown at left below. A puck of mass m, moving at speed v hits an identical puck which is fastened to a pole using a string of length r. After the collision, the puck attached to the string revolves around the pole. Suppose we now lengthen the string by a factor 2, as shown on the right, and repeat the experiment. Compared to the angular speed in the first situation, the new angular speed is

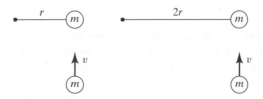

1. twice as high
2. the same

3. half as much

4. none of the above

Answer: 3. If we ignore the effect of the string during the collision between the two pucks, the incoming puck comes to rest and the puck on the string gets a speed v. Because of the string, however, this puck can't move in a straight line, but is instead constrained to moving in a circle. Because its speed is v, it takes a time interval $2\pi r/v$ to complete one revolution around the axis. The angular speed of the puck is $v = \omega r$, and so $\omega = v/r$.

When we lengthen the string by a factor of 2, the circle made by the puck gets twice as large. Because the collision still gives the puck the same speed, it now takes the puck twice as long to complete one revolution. Its angular speed is therefore $\omega = v/2r$.

15. A figure skater stands on one spot on the ice (assumed frictionless) and spins around with her arms extended. When she pulls in her arms, she reduces her rotational inertia and her angular speed increases so that her angular momentum is conserved. Compared to her initial rotational kinetic energy, her rotational kinetic energy after she has pulled in her arms must be

1. the same.

2. larger because she's rotating faster.

3. smaller because her rotational inertia is smaller.

Answer: 2. Rotational kinetic energy is $\frac{1}{2} I\omega^2$. Substituting L, the angular momentum, for $I\omega$, we find $K_{rot} = \frac{1}{2} L\omega$. We know (*a*) that L is constant because the angular momentum is conserved in this isolated system and (*b*) that ω increases. Therefore K_{rot} must increase. The "extra" energy comes from the work she does on her arms: She has to pull them in against their tendency to continue in a straight line, thereby doing work on them. If you prefer to consider the skater and her arms as one system, the extra energy comes from potential energy.

16. Two cylinders of the same size and mass roll down an incline. Cylinder *A* has most of its weight concentrated at the rim, while cylinder *B* has most of its weight concentrated at the center. Which reaches the bottom of the incline first?

1. *A*

2. *B*

3. Both reach the bottom at the same time.

Answer: 2. When the cylinders roll down the incline, gravitational potential energy gets converted to translational and rotational kinetic energy: $mgh = \frac{1}{2} mv^2 + \frac{1}{2} I\omega^2$. We can substitute v/r for ω and write $mgh = \frac{1}{2}(m + I/r^2)v^2$. The values for m and r are the same for both cylinders, and so the cylinder having the smaller rotational inertia has the larger speed. The smaller rotational inertia is obtained when more of the mass is concentrated near the center.

17. A solid disk and a ring roll down an incline. The ring is slower than the disk if

1. $m_{ring} = m_{disk}$, where m is the inertial mass.
2. $r_{ring} = r_{disk}$, where r is the radius.
3. $m_{ring} = m_{disk}$ and $r_{ring} = r_{disk}$.
4. The ring is always slower regardless of the relative values of m and r.

Answer: 4. The ring has more rotational inertia per unit mass than the disk. Therefore as it starts rolling, it has a relatively larger fraction of its total kinetic energy in rotational form and so its translational kinetic energy is lower than that of the disk.

As the disk and ring roll down the incline, the potential energy of each is reduced by an amount mgh, where h is the difference in height between the bottom and top of the incline. This energy is converted to kinetic energy (translational and rotational): $mgh = \frac{1}{2} mv^2 + \frac{1}{2} I\omega^2$. We can write the rotational inertia as $I = cmR^2$, where c is a constant equal to 1 for the ring and $\frac{1}{2}$ for the disk. Because $\omega = v/R$, we have: $mgh = \frac{1}{2} mv^2 + \frac{1}{2} cmR^2(v/R)^2 = \frac{1}{2}(1 + c)mv^2$. The larger c, therefore, the smaller v. Thus the ring, which has the larger rotational inertia, takes longer to go down the incline, regardless of inertia and radius.

18. Two wheels with fixed hubs, each having a mass of 1 kg, start from rest, and forces are applied as shown. Assume the hubs and spokes are massless, so that the rotational inertia is $I = mR^2$. In order to impart identical angular accelerations, how large must F_2 be?

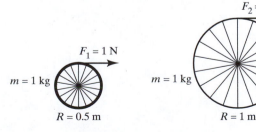

1. 0.25 N

2. 0.5 N

3. 1 N

4. 2 N

5. 4 N

Answer: 4. For each wheel, the moment of inertia times the angular acceleration must equal the force on the wheel rim times the distance from the rim to the center of the wheel.

19. Two wheels initially at rest roll the same distance without slipping down identical inclined planes starting from rest. Wheel B has twice the radius but the same mass as wheel A. All the mass is concentrated in their rims, so that the rotational inertias are $I = mR^2$. Which has more translational kinetic energy when it gets to the bottom?

    1. Wheel A

    2. Wheel B

    3. The kinetic energies are the same.

    4. need more information

    Answer: 3. For each wheel, the gain in total kinetic energy (translational plus rotational) equals the loss in gravitational potential energy. Because the two wheels have identical mass and roll down the same distance they both lose the same amount of potential energy. Both wheels also have the same ratio of translational to rotational kinetic energy, so their translational kinetic energies are the same.

20. Consider the uniformly rotating object shown below. If the object's angular velocity is a vector (in other words, it points in a certain direction in space) is there a particular direction we should associate with the angular velocity?

1. yes, ±*x*

2. yes, ±*y*

3. yes, ±*z*

4. yes, some other direction

5. no, the choice is really arbitrary

Answer: 3. Only the *z* direction—the direction perpendicular to the plane of rotation—is unique; the *x* and *y* directions represent only the instantaneous direction in which the object is moving (*i.e.*, at the instant represented in the diagram). The *z* direction has the same relationship to the instantaneous velocity at all times during the motion.

21. A person spins a tennis ball on a string in a horizontal circle (so that the axis of rotation is vertical). At the point indicated below, the ball is given a sharp blow in the forward direction. This causes a change in angular momentum ΔL in the

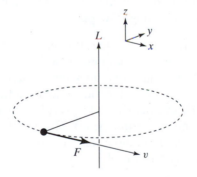

1. *x* direction

2. *y* direction

3. *z* direction

Answer: 3. The force *F* makes the ball go around the circle faster than before, and so the ball's angular momentum increases in magnitude without changing direction. According to the right-hand rule, the angular velocity is in the +*z* direction, and so the change in angular momentum must also be in the +*z* direction.

22. A person spins a tennis ball on a string in a horizontal circle (so that the axis of rotation is vertical). At the point indicated below, the ball is given a sharp blow vertically downward. In which direction does the axis of rotation tilt after the blow?

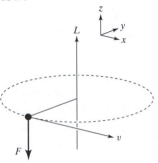

1. +x direction
2. −x direction
3. +y direction
4. −y direction
5. It stays the same (but the magnitude of the angular momentum changes).
6. The ball starts wobbling in all directions.

Answer: 1. As it receives the blow, the ball accelerates downward and its velocity gains a downward component. Still constrained by the string to go around in a circle, the ball moves in a new circular path the plane of which is determined by the instantaneous velocity *v* the ball has immediately after the blow. This velocity *v′* defines a plane that is tilted downward in the forward direction (+x direction). The angular momentum therefore also tilts forward in the +x direction. The change in angular momentum is the dashed vector connecting *L* and *L′*.

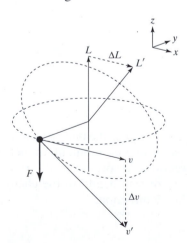

Explanation using torque: According to the right-hand rule, the torque exerted by the force is in the forward direction (+*x*), and so the change in angular momentum must also be in this direction.

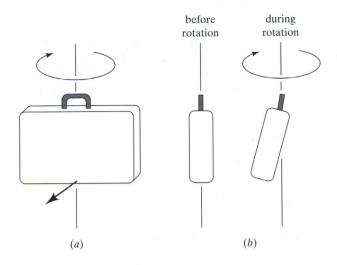

before
rotation

during
rotation

(*a*)

(*b*)

23. A suitcase containing a spinning flywheel is rotated about the vertical axis as shown in (*a*). As it rotates, the bottom of the suitcase moves out and up, as in (*b*). From this, we can conclude that the flywheel, as seen from the side of the suitcase as in (*a*), rotates

1. clockwise.

2. counterclockwise.

Answer: 1. To rotate the suitcase in the direction indicated, a downward torque must be exerted on it. This downward torque gives a downward component to the angular momentum. Because the angular momentum initially points into the page, this downward component causes the suitcase to tilt as indicated. The flywheel must therefore be spinning clockwise.

OSCILLATIONS

1. An object is in equilibrium when the net force and the net torque on it is zero. Which of the following statements is/are correct for an object in an inertial frame of reference?

    A. Any object in equilibrium is at rest.

    B. An object in equilibrium need not be at rest.

    C. An object at rest must be in equilibrium.

    Answer: B. An object is in equilibrium when there is no net force on it. This means the acceleration of the object is zero and hence the object is either at rest or moving at constant velocity. Therefore A is not true, but B is. Choice C is not true. Just think of an object launched up in the air. At the highest point, the object is momentarily at rest ($v = 0$), but the acceleration $a = -g$ is certainly not zero.

2. An object can oscillate around

    1. any equilibrium point.

    2. any stable equilibrium point.

    3. certain stable equilibrium points.

    4. any point, provided the forces exerted on it obey Hooke's law.

    5. any point.

    Answer: 2. When an object in stable equilibrium is disturbed, it tends to return to the equilibrium point. This is the basic requirement for an oscillation. An object in unstable equilibrium tends to move farther away from equilibrium when it is disturbed. Forces around any stable equilibrium obey Hooke's law, provided the displacement from equilibrium is not too large, so any stable equilibrium point will do.

3. Which of the following is necessary to make an object oscillate?

    A. a stable equilibrium

    B. little or no friction

    C. a disturbance

    Answer: All three. A stable equilibrium point is needed. In addition, some sort of disturbance is required to set the object in motion (otherwise the object would simply remain at the equilibrium point). Finally, friction should be small or absent, otherwise the object doesn't oscillate but just returns to the equilibrium point without overshooting it.

4.   A mass attached to a spring oscillates back and forth as indicated in the position vs. time plot below. At point P, the mass has

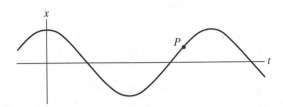

1. positive velocity and positive acceleration.
2. positive velocity and negative acceleration.
3. positive velocity and zero acceleration.
4. negative velocity and positive acceleration.
5. negative velocity and negative acceleration.
6. negative velocity and zero acceleration.
7. zero velocity but is accelerating (positively or negatively).
8. zero velocity and zero acceleration.

Answer: 2. The velocity is positive because the slope of the curve at point P is positive. The acceleration is negative because the curve is concave down at P.

5.   A mass suspended from a spring is oscillating up and down as indicated. Consider two possibilities: (*i*) at some point during the oscillation the mass has zero velocity but is accelerating (positively or negatively); (*ii*) at some point during the oscillation the mass has zero velocity and zero acceleration.

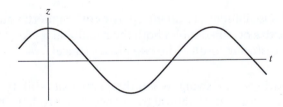

1. Both occur sometime during the oscillation.
2. Neither occurs during the oscillation.
3. Only (*i*) occurs.

4. Only (*ii*) occurs.

Answer: 3. The velocity is zero at the maxima and minima of the curve. At these points, the curve is concave either up or down, and so the particle is accelerating. The particle has zero acceleration at the points of inflection of the curve, which occur at those times *t* for which $x = 0$. At these points, the curve has nonzero slope and so the velocity cannot be zero.

6. An object hangs motionless from a spring. When the object is pulled down, the sum of the elastic potential energy of the spring and the gravitational potential energy of the object and Earth

   1. increases.
   2. stays the same.
   3. decreases.

   Answer: 1. If released from its new position, the object accelerates upward and passes the equilibrium point with nonzero velocity. The object has therefore gained kinetic energy. The two forms of potential energy present are elastic potential energy of the spring and gravitational potential energy. Even though the latter decreases as the object is pulled down, the sum of the two must increase for the object to be able to gain kinetic energy.

7. A person swings on a swing. When the person sits still, the swing oscillates back and forth at its natural frequency. If, instead, two people sit on the swing, the new natural frequency of the swing is

   1. greater.
   2. the same.
   3. smaller.

   Answer: 2. Oscillations are an interplay between inertia and a restoring force. The extra person doubles both the rotational inertia of the swing as well as the restoring torque. The two effects cancel.

8. A person swings on a swing. When the person sits still, the swing oscillates back and forth at its natural frequency. If, instead, the person stands on the swing, the new natural frequency of the swing is

   1. greater.
   2. the same.
   3. smaller.

Answer: 1. By standing up, the distance of the center of mass to the pivot point is reduced. The restoring torque decreases linearly with this distance; the rotational inertia as the square of it. Thus, the rotational inertia decreases more and the period decreases.

9. The traces below show beats that occur when two different pairs of waves are added. For which of the two is the difference in frequency of the original waves greater?

pair 1   pair 2

1. Pair 1
2. Pair 2
3. The frequency difference was the same for both pairs of waves.
4. Need more information.

Answer: 1. The frequency of the envelope of the beat function is proportional to the difference in frequency between the two waves. The greater the difference in frequency, the greater the frequency of the envelope.

10. The traces below show beats that occur when two different pairs of waves are added. Which of the two pairs of original waves contains the wave with the highest frequency?

pair 1   pair 2

1. Pair 1
2. Pair 2
3. The frequency difference was the same for both pairs of waves.
4. Need more information.

Answer: 2. The frequency of the oscillation within the beats is equal to the average of the original frequencies.

11. The circular object pictured here is made to rotate clockwise at 29 revolutions per second. It is filmed with a camera that takes 30 frames per second. Compared to its actual motion, the dot on the film appears to move

1. clockwise at a very slow rate.
2. counterclockwise at a very slow rate.
3. clockwise at a very fast rate.
4. counterclockwise at a very fast rate.
5. in a random fashion.

Answer: 2. After 1/30th of a second, the object has completed 29/30th of a full revolution. To the camera, it therefore appears in position 1—as if it had rotated 1/30th of a turn in the counterclockwise direction. For each subsequent frame, the dot appears to rotate another 1/30th of a rotation in the counterclockwise direction. After 30 frames, the dot has made one complete counterclockwise rotation. As seen by the camera, therefore, the rate of rotation is not 29 clockwise revolutions per second, but 1 counterclockwise revolution per second.

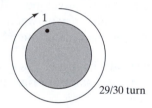

12. A wave pulse is moving, as illustrated, with uniform speed v along a rope. Which of the graphs 1–4 below correctly shows the relation between the displacement s of point P and time t?

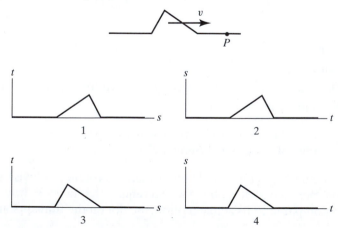

Answer: 2. First the shallow side of the pulse moves through *P*, then the steep side. So the displacement increases more slowly than it decreases. Note that diagram 1 and 3 do not represent a physically possible situation: In both instances, the diagram indicates more than one possible value of displacement at a given instant.

13. A wave is sent along a long spring by moving the left end rapidly to the right and keeping it there. The figure shows the wave pulse at *QR*—part *RS* of the long spring is as yet undisturbed. Which of the graphs 1–5 correctly shows the relation between displacement *s* and position *x*? (Displacements to the right are positive.)

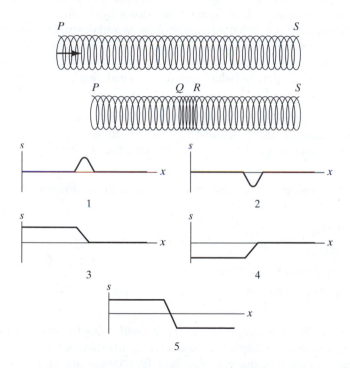

Answer: 3. The left end of the long spring (point *P*) is displaced toward the right and kept there. The displacement *s* of point *P* is therefore nonzero. Only graphs 3 and 5 show a nonzero displacement at the left. At the instant shown, the wavepulse has not yet reached the right end of the long spring (point *S*), so this end has not yet been displaced. Graph 5 shows a negative displacement at the right end, so the only possible choice is graph 3.

14. Two strings, one thick and the other thin, are connected to form one long string. A wave travels along the string and passes the point where the two strings are connected. Which of the following change(s) at that point:

A. frequency

B. period

C. propagation speed

D. wavelength

Answer: C and D. First consider the propagation of the wave along the first string. As the wave travels along the string, its shape and propagation speed are constant. So if the pulse is created by shaking one end up and down at a certain frequency, then the other end of that string will shake up and down at the same frequency (but a little later). So the beginning end of the second string is shaken up and down by the first string at precisely the frequency at which it is being shaken up and down. The frequency must therefore remain the same—it is determined by whomever creates the wave by shaking an end of the rope up and down. If the frequency is the same, then the period (which equals $1/f$) must also be the same.

The propagation speed, however, depends on properties of the string (tension, density). Since the two strings are unequal, the speed therefore changes and so, too, the wavelength, since the wavelength is given by v/f.

15. By shaking one end of a stretched string, a single pulse is generated. The traveling pulse carries

1. energy.

2. momentum.

3. energy and momentum.

4. neither of the two.

Answer: 3. A traveling wave does not involve any transport of mass (even though mass gets displaced, no mass actually travels along with the wave from one point in space to another). By shaking one end of a rope, however, one can get the other end to move, and when that end is put in motion, it has both momentum and energy. So when the wave travels down the rope, it carries both momentum and energy.

16. A weight is hung over a pulley and attached to a string composed of two parts, each made of the same material but one having four times the diameter of the other. The string is plucked so that a pulse moves along it, moving at speed v_1 in the thick part and at speed v_2 in the thin part. What is v_1/v_2?

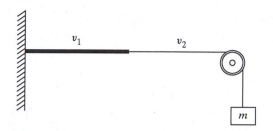

1. 1
2. 2
3. 1/2
4. 1/4

Answer: 4. The tension is constant along the string, but the linear mass density is 16 times greater in the thicker part. The speed of a traveling wave varies inversely with the square root of this density.

17. Two identical symmetric pulses of opposite amplitude travel along a stretched string and interfere destructively. Which of the following is/are true?

A. There is an instant at which the string is completely straight.

B. When the two pulses interfere, the energy of the pulses is momentarily zero.

C. There is a point on the string that does not move up or down.

D. There are several points on the string that do not move up or down.

Answer: A and C. Choice A is correct because, when the two pulses are in the same position, they cancel each other exactly, leaving the string completely straight. The point right in the middle between the two pulses can't move—whatever the displacement due to one pulse at that point, it is canceled by the displacement due to the other.

18. A string is clamped at both ends and plucked so it vibrates in a standing mode between two extreme positions *a* and *b*. Let upward motion correspond to positive velocities. When the string is in position *c*, the instantaneous velocity of points along the string:

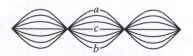

1. is zero everywhere.
2. is positive everywhere.
3. is negative everywhere.
4. depends on location.

Answer: 4. The drawing below shows the string just before (top) and just after (bottom) it reaches position *c* (middle). Notice that dot *P* moves down and dot *R* moves up, whereas dot *Q* doesn't move at all. So the velocities of points around dot *P* are negative and those of points around dot *R* are positive. Dot *Q* doesn't move and has zero velocity at all times.

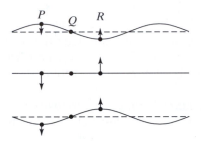

19. A string is clamped at both ends and plucked so it vibrates in a standing mode between two extreme positions *a* and *b*. Let upward motion correspond to positive velocities. When the string is in position *b*, the instantaneous velocity of points along the string

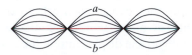

1. is zero everywhere.
2. is positive everywhere.
3. is negative everywhere.
4. depends on location.

Answer: 1. The following drawing shows the string just before (top) and just after (bottom) it reaches position *b* (middle). Dot *R* moves downward just before the string reaches position *b* and upward just afterwards.

The instantaneous velocity of the dot right at b, therefore, is zero. Similarly, the instantaneous velocity of dot P is zero and since dot Q doesn't move at all, its instantaneous velocity, too, is zero.

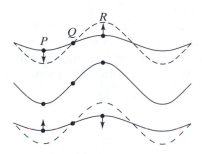

SOUND

1. The four figures below represent sound waves emitted by a moving source. Which picture(s) represent(s) a source moving at less than the speed of sound?

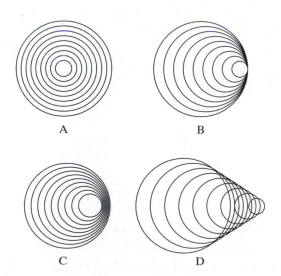

Answer: A and C. In figure B, (see following figure) the wavefront created when the source was at the center of circle 1 has traveled the distance between lines 1 and 3. Similarly, the wavefront created when the source was at the center of circle 2 has traveled the distance between lines 2 and

3. This means that the source had to move at exactly the speed of the sound waves—else the wave emitted at 2 would not catch up with that emitted at 1. The result of this motion at the speed of sound is a piling up of sound waves in front of the source at 3. In figure D, the wavefronts emitted later have moved beyond those emitted earlier, indicating that the source moves faster than the speed of sound. In figure A, all the wavefronts are concentric, indicating that the source is stationary. Figure C shows a moving source of sound, but the displacement of the wavefronts is less than in B so the source is moving at a speed slower than the speed of sound.

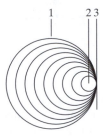

2. Three observers, A, B, and C are listening to a moving source of sound. The diagram below shows the location of the wavecrests of the moving source with respect to the three observers. Which of the following is true?

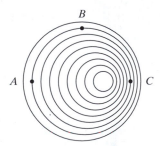

1. The wavefronts move faster at A than at B and C.
2. The wavefronts move faster at C than at A and B.
3. The frequency of the sound is highest at A.
4. The frequency of the sound is highest at B.
5. The frequency of the sound is highest at C.

Answer: 5. The speed at which the wavefronts move is the speed of sound in air, which is independent of the speed of the source and the location of the observers. So the first two choices are incorrect. The observed frequency, however, is determined by the number of wavefronts passing the observer per unit time. So, the more closely the wavefronts are spaced, the higher the frequency. Inspection of the figure shows that the wavefronts are most closely spaced for observer *C*.

3. The following figure shows the wavefronts generated by an airplane flying past an observer *A* at a speed greater than that of sound. After the airplane has passed, the observer reports hearing

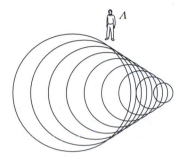

1. a sonic boom only when the airplane breaks the sound barrier, then nothing.
2. a succession of sonic booms.
3. a sonic boom, then silence.
4. first nothing, then a sonic boom, then the sound of engines.
5. no sonic boom because the airplane flew faster than sound all along.

Answer: 4. The observer first hears nothing because the sound waves haven't reached the location of the observer yet. When the wake caused by the sound waves reaches the observer, a sonic boom is caused by the overlapping crests of the sound waves. After this, the sound of the engines is heard, as indicated by the regularly spaced sound waves inside the wake.

FLUID STATICS

1. Imagine holding two identical bricks under water. Brick *A* is just beneath the surface of the water, while brick *B* is at a greater depth. The force needed to hold brick *B* in place is
   1. larger
   2. the same as

3. smaller

than the force required to hold brick *A* in place.

Answer: 2. The buoyant force on each brick is equal to the weight of water it displaces and does not depend on depth.

2.   When a hole is made in the side of a container holding water, water flows out and follows a parabolic trajectory. If the container is dropped in free fall, the water flow

1. diminishes.

2. stops altogether.

3. goes out in a straight line.

4. curves upward.

Answer: 2. When the container is at rest with respect to Earth, there is pressure on the walls of the container due to the water. The pressure depends on the depth and is equal to $\rho g h$, with ρ being the density of water. When the container is in free fall, both the water and the container have an acceleration of zero, not *g*, in the container frame of reference. In this frame, the pressure of the water on the walls of the container is zero, so there is no outward flow.

3.   A container is filled with oil and fitted on both ends with pistons. The area of the left piston is 10 mm^2; that of the right piston 10,000 mm^2. What force must be exerted on the left piston to keep the 10,000-N car on the right at the same height?

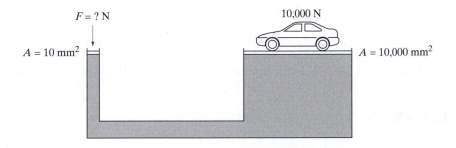

1. 10 N

2. 100 N

3. 10,000 N

4. 10^6 N

5. 10^8 N

6. insufficient information

Answer: 1. To keep the fluid levels equal, equal pressure (force per unit area) must be exerted on the two pistons.

4. A 200-ton ship enters the lock of a canal. The fit between the sides of the lock and the ship is tight so that the weight of the water left in the lock after it closes is much less than 200 tons. Can the ship still float if the quantity of water left in the lock is much less than the ship's weight?

1. Yes, as long as the water gets up to the ship's waterline.

2. No, the ship touches bottom because it weighs more than the water in the lock.

Answer: 1. What matters is not the weight of the water left in the lock, but the weight of the water forced out of the lock by the ship. As long as the density of the ship is less than that of water and the water gets to the waterline, it floats.

5. Two identical glasses are filled to the same level with water. One of the two glasses has ice cubes floating in it. Which weighs more?

1. The glass without ice cubes.

2. The glass with ice cubes.

3. The two weigh the same.

Answer: 3. The ice cubes displace exactly their own weight in water, so the two glasses weigh the same amount.

6. Two identical glasses are filled to the same level with water. One of the two glasses has ice cubes floating in it. When the ice cubes melt, in which glass is the level of the water higher?

1. The glass without ice cubes.

2. The glass with ice cubes.

3. It is the same in both.

Answer: 3. When the ice cubes melt, they turn into an amount of water having the same weight. This weight is also equal to the weight of the water originally displaced by the cubes. Since the densities of the melted ice and the surrounding water are identical, the volume occupied by the melted ice exactly equals the volume of displaced water.

7. Two cups are filled to the same level with water. One of the two cups has plastic balls floating in it. If the density of the plastic balls is less than that of ice, which of the two cups weighs more?

    1. The cup without plastic balls.
    2. The cup with plastic balls.
    3. The two weigh the same.

    Answer: 3. The plastic balls displace exactly their own weight in water, so the two glasses weigh the same amount.

8. A lead weight is fastened on top of a large solid piece of Styrofoam that floats in a container of water. Because of the weight of the lead, the water line is flush with the top surface of the Styrofoam. If the piece of Styrofoam is turned upside down so that the weight is now suspended underneath it,

    1. the arrangement sinks.
    2. the water line is below the top surface of the Styrofoam.
    3. the water line is still flush with the top surface of the Styrofoam.

    Answer: 2. Since both the weight and the Styrofoam are solid, the arrangement floats in both orientations. Therefore, turning the arrangement upside down does not change the buoyant force; however, since the weight is now submerged, it displaces a volume of water. To maintain the same buoyant force, the Styrofoam must therefore displace less water than before.

9. A lead weight is fastened to a large solid piece of Styrofoam that floats in a container of water. Because of the weight of the lead, the water line is flush with the top surface of the Styrofoam. If the piece of Styrofoam is turned upside down, so that the weight is now suspended underneath it, the water level in the container

    1. rises.
    2. drops.
    3. remains the same.

    Answer: 3. The buoyant force is unchanged, so the volume of water displaced is unchanged and the water level stays the same.

10. A boat carrying a large boulder is floating on a lake. The boulder is thrown overboard and sinks. The water level in the lake (with respect to the shore)

    1. rises.
    2. drops.
    3. remains the same.

Answer: 2. When it is inside the boat, the boulder displaces its weight in water. When it is thrown overboard, it only displaces its volume in water so the water level of the lake with respect to the shore goes down.

11. Consider an object that floats in water but sinks in oil. When the object floats in water, half of it is submerged. If we slowly pour oil on top of the water so it completely covers the object, the object

1. moves up.

2. stays in the same place.

3. moves down.

Answer: 1. With the oil atop the water, there is an additional buoyant force on the object equal to the weight of the displaced oil. The effect of this additional force is to displace the object upward.

12. Consider an object floating in a container of water. If the container is placed in an elevator that accelerates upward,

1. more of the object is below water.

2. less of the object is below water.

3. there is no difference.

Answer: 3. The acceleration effectively increases g, changing the apparent weight of objects inside the elevator. However, there is no difference in the amount that the object is submerged because the acceleration affects the water and the object in the same amount.

13. A circular hoop sits in a stream of water, oriented perpendicular to the current. If the area of the hoop is doubled, the flux (volume of water per unit time) through it

1. decreases by a factor of 4.

2. decreases by a factor of 2.

3. remains the same.

4. increases by a factor of 2.

5. increases by a factor of 4.

Answer: 4. The flux is equal to the flow velocity times the cross-sectional area through which the fluid flows. If the area is doubled, so is the flux.

14. Blood flows through a coronary artery that is partially blocked by deposits along the artery wall. Through which part of the artery is the flux (volume of blood per unit time) largest?

1. The narrow part.
2. The wide part.
3. The flux is the same in both parts.

Answer: 3. Because liquids, such as blood, are incompressible and because no blood accumulates in (or leaks out of) the artery, the flux is the same everywhere.

15. Blood flows through a coronary artery that is partially blocked by deposits along the artery wall. Through which part of the artery is the flow speed largest?

1. The narrow part.
2. The wide part.
3. The flow speed is the same in both parts.

Answer: 1. Because liquids, such as blood, are incompressible and because no blood accumulates in (or leaks out of) the artery, the flux is the same everywhere. Since the flux is equal to the flow speed times the cross-sectional area of the artery, the blood must flow faster in the narrow part.

16. Two hoses, one of 20-mm diameter, the other of 15-mm diameter are connected one behind the other to a faucet. At the open end of the hose, the flow of water measures 10 liters per minute. Through which pipe does the water flow faster?

1. the 20-mm hose
2. the 15-mm hose
3. The flow rate is the same in both cases.
4. The answer depends on which of the two hoses comes first in the flow.

Answer: 2. Unless the hose is leaking, the flux, or volume of fluid per unit time through each hose is the same. Since the flux is equal to the flow speed times the cross-sectional area of the hose, the water must flow faster in the narrower hose.

17. A blood platelet drifts along with the flow of blood through an artery that is partially blocked by deposits. As the platelet moves from the narrow region to the wider region, its speed

   1. increases.
   2. remains the same.
   3. decreases.

Answer: 3. Because liquids, such as blood, are incompressible and because no blood accumulates in (or leaks out of) the artery, the flux is the same everywhere. Since the flux is equal to the flow speed times the cross-sectional area of the artery, the blood must flow more slowly in the wide part than it does in the narrow part. Hence the platelet slows down as it enters the wider part.

18. A blood platelet drifts along with the flow of blood through an artery that is partially blocked by deposits. As the platelet moves from the narrow region to the wider region, it experiences

   1. an increase in pressure.
   2. no change in pressure.
   3. a decrease in pressure.

Answer: 1. Because liquids, such as blood, are incompressible and because no blood accumulates in (or leaks out of) the artery, the flux is the same everywhere. Since the flux is equal to the flow speed times the cross-sectional area of the artery, the blood must flow more slowly in the wide part than it does in the narrow part. Hence the platelet slows down as it enters the wider part. This implies that the surrounding liquid exerts a force on the platelet in a direction opposite to its travel. The pressure ahead of the platelet must therefore be larger than that behind it.

OPTICS

1. A group of sprinters gather at point P on a parking lot bordering a beach. They must run across the parking lot to a point Q on the beach as quickly as possible. Which path from P to Q takes the least time? You should consider the relative speeds of the sprinters on the hard surface of the parking lot and on loose sand.

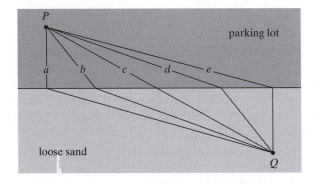

1. a
2. b
3. c
4. d
5. e
6. All paths take the same amount of time.

Answer: 4. Anybody—sprinter or couch potato—can run more quickly on a hard surface than on loose sand. While the distance on loose sand is slightly less for path e than for path d, the run over the parking lot is much longer. The result is that path e is more time-consuming than path d.

2. Suppose the sprinters wish to get from point Q on the beach to point P on the parking lot as quickly as possible. Which path takes the least time?

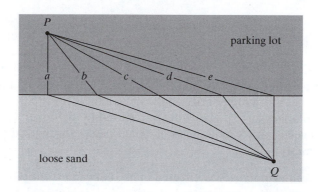

1. *a*
2. *b*
3. *c*
4. *d*
5. *e*
6. All paths take the same amount of time.

Answer: 4. Anybody—sprinter or couch potato—can run more quickly on a hard surface than on loose sand. While the distance on loose sand is slightly less for path *e* than for path *d*, the run over the parking lot is much longer. The result is that path *e* is more time-consuming than path *d*.

3. When a ray of light is incident on two polarizers with their polarization axes perpendicular, no light is transmitted. If a third polarizer is inserted between these two with its polarization axis at 45° to that of the other two, does any light get through to point *P*?

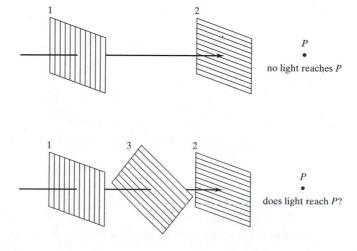

1. yes

2. no

Answer: 1. The light transmitted by the first polarizer has its polarization vector oriented vertically. So, while no light is transmitted through a polarizer that is oriented horizontally, some light is transmitted through the third polarizer at 45°. The light transmitted by polarizer 3 has its polarization vector at 45°, and thus some light passes through (the horizontally oriented) polarizer 2.

4. When a third polarizer is inserted at 45° between two orthogonal polarizers, some light is transmitted. If, instead of a single polarizer at 45°, we insert a large number N of polarizers, each time rotating the axis of polarization over an angle 90°/N,

    1. no light

    2. less light

    3. the same amount of light

    4. more light

    gets through.

    Answer: 4. The smaller the angle between two successive polarizers, the larger the amount of transmitted light.

5. An observer O, facing a mirror, observes a light source S. Where does O perceive the mirror image of S to be located?

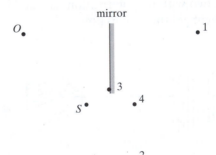

    1. 1

    2. 2

    3. 3

    4. 4

    5. Some other location.

    6. The image of S cannot be seen by O when O and S are located as shown.

Answer: 4. The image of a point in a mirror is always on a line perpendicular to the mirror surface and as far behind the mirror as the point is in front of it.

6. The observer at O views two closely spaced lines through an angled piece of plastic. To the observer, the lines appear (choose all that apply)

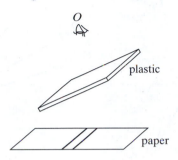

A. shifted to the right.
B. shifted to the left.
C. spaced farther apart.
D. spaced closer together.
E. exactly as they do without the piece of plastic.

Answer: B. Refraction through the plastic shifts the beams toward the left. Since the beams from both lines shift the same amount, the spacing between the two lines remains the same.

7. A fish swims below the surface of the water at P. An observer at O sees the fish at

1. a greater depth than it really is.
2. the same depth.
3. a smaller depth than it really is.

Answer: 3. The rays emerging from the water surface converge to a point above the fish. See figure.

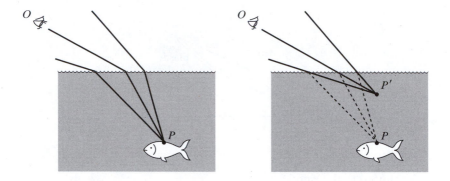

8.   A fish swims below the surface of the water. Suppose an observer is looking at the fish from point O'—straight above the fish. The observer sees the fish at

1. a greater depth than it really is.
2. the same depth.
3. a smaller depth than it really is.

Answer: 3. The rays emerging from the water surface converge to a point above the fish.

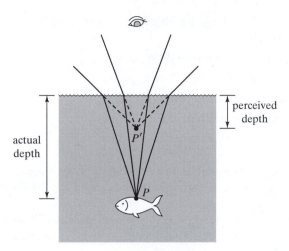

9. A parallel beam of light is sent through an aquarium. If a convex glass lens is held in the water, it focuses the beam

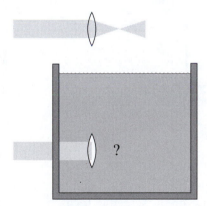

1. closer to the lens than
2. at the same position as
3. farther from the lens than
outside the water.

Answer: 3. The index of refraction of water is between those of air and glass. Thus, rays of light refract less in going from water to glass than in going from air to glass. As a result, a convex lens focuses rays of light less tightly under water.

10.   A lens is used to image an object onto a screen. If the right half of the lens is covered,

1. the left half of the image disappears.
2. the right half of the image disappears.
3. the entire image disappears.
4. the image becomes blurred.
5. the image becomes fainter.

Answer: 5. Any uncovered part of the lens forms a complete image. Since only half the light gets through the lens, however, the image is fainter.

11.   The lens in an overhead projector forms an image P' of a point P on an overhead transparency. If the screen is moved closer to the projector, the lens must be:

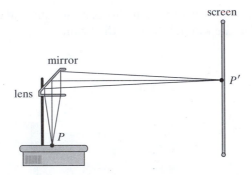

1. moved up
2. left in place
3. moved down

to keep the image on the screen in focus.

Answer: 1. If the screen is moved closer, the rays need to be focused more strongly. Since the angle of refraction of a ray of light incident near the edge of a lens is fixed, the incident rays must be less divergent. To accomplish this, the lens must be moved up.

12.   Monochromatic light shines on a pair of identical glass microscope slides that form a very narrow wedge. The top surface of the upper slide and the bottom surface of the lower slide have special coatings on them so that they reflect no light. The inner two surfaces (A and B) have nonzero reflectivities. A top view of the slides looks like

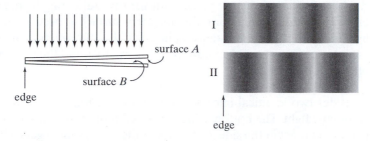

1. I.
2. II.

Answer: 1. The two inner surfaces produce an interference pattern that varies from dark to light as the separation between the slides increases. Light reflected from the surface *A* undergoes no phase shift, while light reflected from surface *B* undergoes a phase shift of 180°. Where the surfaces have near zero separation, we see total destructive interference. The resulting pattern has a dark band near the point of contact of the two slides.

13. Monochromatic light shines on a pair of microscope slides that form a very narrow wedge. The top slide is made of crown glass ($n = 1.5$) and the bottom slide of flint glass ($n = 1.7$). Both slides are immersed in sassafras oil, which has an index intermediate between those of the two slides. The top surface of the upper slide and the bottom surface of the lower slide have special coatings on them so that they reflect no light. The inner two surfaces (*A* and *B*) have nonzero reflectivities. A top view of the slides looks like

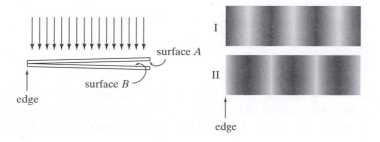

1. I.
2. II.

Answer: 2. There is a 180° phase shift for rays reflecting from both surfaces. The phase change for reflection from surface *A* occurs because the light is coming from a substance of higher index of refraction. In this case, the interference at zero separation is constructive, and the pattern begins off with a bright band.

14. Consider two identical microscope slides in air illuminated with monochromatic light. The bottom slide is rotated (counterclockwise about the point of contact in the side view) so that the wedge angle gets a bit smaller. What happens to the fringes?

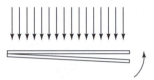

   1. They are spaced farther apart.
   2. They are spaced closer together.
   3. They don't change.

   Answer: 1. Total constructive interference occurs at points where the separation between the slides is equal to an odd half-integer multiple of the wavelength. As the angle becomes smaller, these points move farther apart.

15. Two identical slides in air are illuminated with monochromatic light. The slides are exactly parallel, and the top slide is moving slowly upward. What do you see in top view?
   1. all black
   2. all bright
   3. fringes moving apart
   4. sequentially all black, then all bright
   5. none of the above

   Answer: 4. Light reflected from the top slide interferes with light reflected from the bottom slide. As the separation between the two slides varies, the interference varies between totally constructive and totally destructive.

16. Diffraction occurs when light passes a:
   A. pinhole.
   B. narrow slit.

C. wide slit.

D. sharp edge.

E. all of the above

Answer: E. Any time a beam of light is clipped by an edge, it diffracts. The amount of diffraction depends on the ratio of the wavelength of the light to the size of the opening through which light is transmitted.

17. The Huygens-Fresnel principle tells us to pretend that each point of a wavefront in a slit or aperture is a point source of light emitting a spherical wave. Is this true only for points inside the slit? What if there is no slit? The Huygens-Fresnel principle really applies

    1. to any point anywhere in a beam path.
    2. to any point in a beam path where matter is present.
    3. only in slits or apertures.

    Answer: 1. The principle treats all points in the beam path as point sources of light emitting spherical waves.

18. If the Huygens-Fresnel principle applies to any point anywhere in a beam path, why doesn't a laser beam without any slit spread out in all directions?

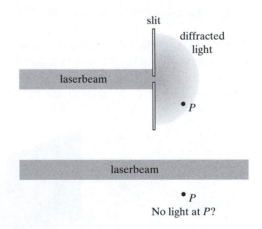

1. Because all waves that spread interfere destructively.
2. It does spread, but the spread is so small that we normally don't notice it.
3. We can't apply the Huygens-Fresnel principle anywhere but in slits and apertures.

Answer: 1. Those light waves from the laser that do spread, interfere destructively. In the presence of a narrow slit, there are not enough rays to interfere destructively.

19. Imagine holding a circular disk in a beam of monochromatic light. If diffraction occurs at the edge of the disk, the center of the shadow of the disk is

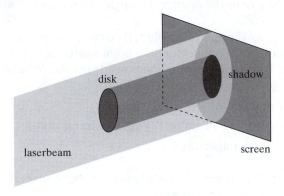

1. a bright spot.
2. darker than the rest of the shadow.
3. bright or dark, depending on the distance between the disk and the screen.
4. as dark as the rest of the shadow, but less dark than if there is no diffraction.

Answer: 1. All the rays from the edge of the disk add in phase at the center of the shadow. The resulting bright spot is called the Poisson spot.

20. The pattern on the screen is due to a narrow slit that is

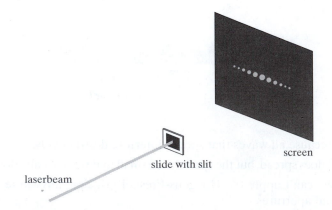

1. horizontal.
2. vertical.

Answer: 2. Diffraction is most pronounced for small apertures, and hence diffraction occurs in the direction of the smallest dimension of the slit.

21. Suppose we cover each slit in Young's experiment with a polarizer such that the polarization transmitted by each slit is orthogonal to that transmitted through the other. On a screen behind the slits, we see:
    1. the usual fringe pattern.
    2. the usual fringes shifted over such that the maxima occur where the minima used to be.
    3. nothing at all.
    4. a fairly uniformly illuminated elongated spot.

    Answer: 4. Because of the orthogonal polarization, light from the two slits does not interfere, but simply adds. Because each slit causes diffraction, the result is an elongated spot on the screen.

22. A diffraction grating is illuminated with yellow light at normal incidence. The pattern seen on a screen behind the grating consists of three yellow spots, one at zero degrees (straight through) and one each at ±45°. You now add red light of equal intensity, coming in the same direction as the yellow light. The new pattern consists of
    1. red spots at 0° and ±45°.
    2. yellow spots at 0° and ±45°.
    3. orange spots at 0° and ±45°.
    4. an orange spot at 0°, yellow spots at ±45°, and red spots slightly farther out.
    5. an orange spot at 0°, yellow spots at ±45°, and red spots slightly closer in.

    Answer: 4. Because there is always a central maximum for each wavelength, there must be an (red + yellow =) orange spot at 0°. The distance between peaks of the subsequent maxima varies directly with wavelength, and so the red spots are farther out than the yellow.

23. A planar wave is incident on a pair of slits whose width and separation are comparable to the wavelength of the incident wave. Seen on a screen behind the slits is/are
    1. two spots, one behind each slit.
    2. only one spot, behind the center of the pair of slits.
    3. many spots distributed randomly.
    4. many spots distributed evenly.

Answer: 4. The diffraction pattern for a pair of slits is not random but consists of many maxima and minima of intensity.

24. An interference pattern is formed on a screen by shining a planar wave on a double-slit arrangement (left). If we cover one slit with a glass plate (right), the phases of the two emerging waves will be different because the wavelength is shorter in glass than in air. If the phase difference is 180°, how is the interference pattern, shown left, altered?

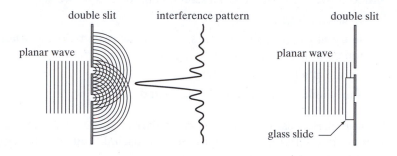

1. The pattern vanishes.
2. The bright spots lie closer together.
3. The bright spots are farther apart.
4. There are no changes.
5. Bright and dark spots are interchanged.

Answer: 5. The diffraction pattern results from the interference of light emerging from the two slits. Introducing a 180° phase shift into the light from one of the slits changes what would have been a point of constructive interference to a point of destructive interference, and vice versa.

25. For a given lens diameter, which light gives the best resolution in a microscope?
1. red
2. yellow
3. green
4. blue
5. All give the same resolution.

Answer: 4. Diffraction is proportional to the aperture divided by the wavelength of the light. If the aperture, determined by the diameter of the lens, is fixed, resolution is inversely proportional to wavelength. Thus, blue light, having the shortest wavelength, gives the best resolution.

26. Blue light of wavelength λ passes through a single slit of width a and forms a diffraction pattern on a screen. If the blue light is replaced by red light of wavelength 2λ, the original diffraction pattern is reproduced if the slit width is changed to

    1. $a/4$.

    2. $a/2$.

    3. No change is necessary.

    4. $2a$.

    5. $4a$.

    6. There is no width that can be used to reproduce the original pattern.

    Answer: 4. The position of the maxima and minima depend on the ratio of slit width to wavelength. To reproduce the original pattern when the wavelength is doubled, one must double the slit width.

27. A planar wave is incident on a single slit and forms a diffraction pattern on a screen. The pattern has a central, zeroth-order maximum and a number of secondary maxima. The first-order maximum is formed in a direction where light from the top third (a) of the slit cancels light from the middle third (b). The intensity of the first-order maximum is thus due only to the bottom third of the light through the slit (c) and is therefore roughly 1/9 of the intensity I of the central maximum. What is the intensity of the second-order maximum?

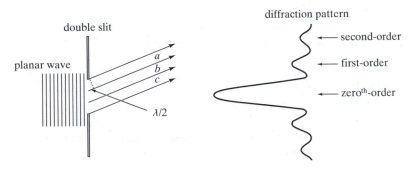

    1. $I/4$

    2. $I/9$

    3. $I/16$

    4. $I/25$

    Answer: 4. The second-order maximum is formed when light from the top fifth cancels the second fifth of the light passing through the slit, as does light from the third and fourth fifth. The second-order maximum is thus due to the bottom fifth of the light passing through the slit. Because the intensity of light is proportional to the square of the electric field, the intensity of the second-order maximum is roughly 1/25 of the incident intensity.

ELECTROSTATICS

1. A positively charged object is placed close to a conducting object attached to an insulating glass pedestal (*a*). After the opposite side of the conductor is grounded for a short time interval (*b*), the conductor becomes negatively charged (*c*). Based on this information, we can conclude that within the conductor

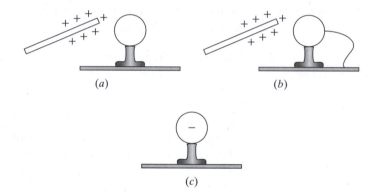

(*a*)   (*b*)

(*c*)

1. both positive and negative charges move freely.
2. only negative charges move freely.
3. only positive charges move freely.
4. We can't really conclude anything.

Answer: 4. The same result is achieved regardless of whether the charge carriers are positive or negative.

2. Three pithballs are suspended from thin threads. Various objects are then rubbed against other objects (nylon against silk, glass against polyester, etc.) and each of the pithballs is charged by touching them with one of these objects. It is found that pithballs 1 and 2 repel each other and that pithballs 2 and 3 repel each other. From this we can conclude that
    1. 1 and 3 carry charges of opposite sign.
    2. 1 and 3 carry charges of equal sign.
    3. all three carry the charges of the same sign.

4. one of the objects carries no charge.

5. we need to do more experiments to determine the sign of the charges.

Answer: 3. Charges of equal sign repel, so 1 and 2 carry charges of equal sign and so, too, do 2 and 3.

3. Three pithballs are suspended from thin threads. Various objects are then rubbed against other objects (nylon against silk, glass against polyester, etc.) and each of the pithballs is charged by touching them with one of these objects. It is found that pithballs 1 and 2 attract each other and that pithballs 2 and 3 repel each other. From this we can conclude that

1. 1 and 3 carry charges of opposite sign.

2. 1 and 3 carry charges of equal sign.

3. all three carry the charges of the same sign.

4. one of the objects carries no charge.

5. we need to do more experiments to determine the sign of the charges.

Answer: 5. Charges of opposite sign attract, charges of equal sign repel, and any type of charge attracts a neutral object. So 1 and 2 either carry charges of opposite sign or one of the two is neutral and the other charged. Since 2 and 3 repel, however, we know that 2 and 3 carry charges of equal sign. So there are two possibilities: 2 and 3 carry charges of equal sign and (*i*) 1 is neutral, or (*ii*) 1 carries a charge of the opposite sign to that of 2 and 3.

4. A hydrogen atom is composed of a nucleus containing a single proton, about which a single electron orbits. The electric force between the two particles is 2.3×10^{39} greater than the gravitational force! If we can adjust the distance between the two particles, can we find a separation at which the electric and gravitational forces are equal?

1. Yes, we must move the particles farther apart.

2. Yes, we must move the particles closer together.

3. no, at any distance

Answer: 3. Both the electric and gravitational forces vary as the inverse square of the separation between two bodies. Thus, the forces cannot be equal at any distance.

5. Two uniformly charged spheres are firmly fastened to and electrically insulated from frictionless pucks on an air table. The charge on sphere 2 is three times the charge on sphere 1. Which force diagram correctly shows the magnitude and direction of the electrostatic forces:

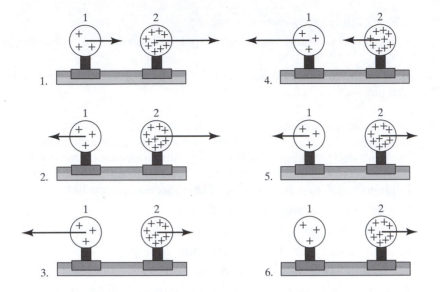

7. none of the above

Answer: 5. The magnitude of the electrostatic force exerted by 2 on 1 is equal to the magnitude of the electrostatic force exerted by 1 on 2. If the charges are of the same sign, the forces are repulsive; if the charges are of opposite sign, the forces are attractive.

6. Consider the four field patterns shown. Assuming there are no charges in the regions shown, which of the patterns represent(s) a possible electrostatic field:

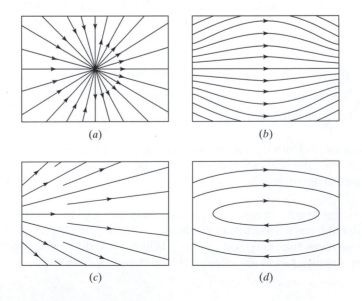

1. (*a*)
2. (*b*)
3. (*b*) and (*d*)
4. (*a*) and (*c*)
5. (*b*) and (*c*)
6. some other combination
7. None of the above.

Answer: 2. Pattern (*a*) can be eliminated because field lines cannot simultaneously emanate from and converge at a single point; (*c*) can be eliminated because there are no charges in the region, and so there are no sources of field lines; (*d*) can be eliminated because electrostatic field lines do not close on themselves.

7. An electrically neutral dipole is placed in an external field. In which situation(s) is the net force on the dipole zero?

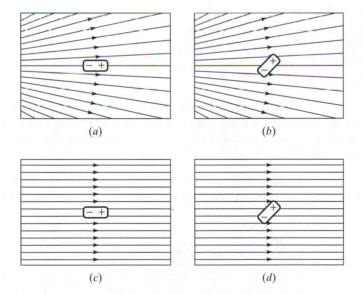

(*a*)   (*b*)

(*c*)   (*d*)

1. (*a*)
2. (*c*)
3. (*b*) and (*d*)
4. (*a*) and (*c*)
5. (*c*) and (*d*)
6. some other combination

7. none of the above

Answer: 5. An electric dipole in a uniform electric field experiences zero net force, and the field is uniform in (c) and (d). Notice that there is a net torque in case (d).

8. The electric charge on plate 1 is $+\sigma$ for plate 1 and on plate 2 is $-\sigma$. The area of both plates is A. The magnitude of the electric field associated with plate 1 is $\sigma/2A\varepsilon_0$, and the electric field lines are as shown. When the two are placed parallel to one another, the magnitude of the electric field is

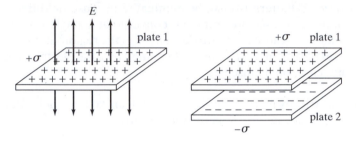

1. $\sigma/A\varepsilon_0$ between, 0 outside.
2. $\sigma/A\varepsilon_0$ between, $\pm\sigma/2A\varepsilon_0$ outside.
3. zero both between and outside.
4. $\pm\sigma/2A\varepsilon_0$ both between and outside.
5. none of the above.

Answer: 1. The magnitude of the electric field for plate 2 is also $\sigma/2A\varepsilon_0$, but the field lines for this plate are directed toward the plate. Using the principle of superposition in the regions between and outside the plates gives answer 1.

9. A cylindrical piece of insulating material is placed in an external electric field, as shown. The net electric flux passing through the surface of the cylinder is

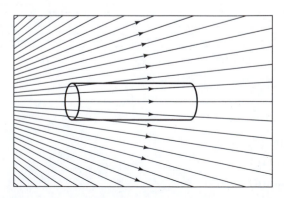

1. positive.
2. negative.
3. zero.

Answer: 3. Because each field line entering the surface also leaves the surface, there is no net flux through it.

10. Two test charges are brought separately into the vicinity of a charge $+Q$. First, test charge $+q$ is brought to point A a distance r from $+Q$. Next, $+q$ is removed and a test charge $+2q$ is brought to point B a distance $2r$ from $+Q$. Compared with the electrostatic potential of the charge at A, that of the charge at B is

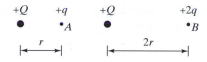

1. greater.
2. smaller.
3. the same.

Answer: 2. The electrostatic potential at A is Q/r, whereas at point B it is $Q/2r$. The magnitudes of the charges q and $2q$ at A and B do not enter into the expression for the electrostatic potential.

11. Two test charges are brought separately into the vicinity of a charge $+Q$. First, test charge $+q$ is brought to a point a distance r from $+Q$. Then this charge is removed and test charge $-q$ is brought to the same point. The electrostatic potential energy of which test charge is greater:

$+Q$ $+q$ $+Q$ $-q$

$\leftarrow r \rightarrow$ $\leftarrow r \rightarrow$

1. $+q$
2. $-q$
3. It is the same for both.

Answer: 1. The electrostatic potential energy of a test charge depends on both its position and its charge. For the $+q$ charge, the electrostatic potential energy equals Qq/r, whereas for the $-q$ charge it equals $-Qq/r$.

12. An electron is pushed into an electric field where it acquires a 1-V electrical potential. Suppose instead that two electrons are pushed the same distance into the same electric field. The electrical potential of the two electrons is

1. 0.25 V.

2. 0.5 V.

3. 1 V.

4. 2 V.

5. 4 V.

Answer: 3. The electrostatic potential of a charged particle in an electric field depends only on the position of the particle, not on its charge.

13. A solid spherical conductor is given a net nonzero charge. The electrostatic potential of the conductor is

1. largest at the center.

2. largest on the surface.

3. largest somewhere between center and surface.

4. constant throughout the volume.

Answer: 4. By definition, the electric field inside a conductor is zero. The electric field is the gradient of the electrostatic potential. Thus, the electrostatic potential inside a conductor must be constant and, by continuity, must be equal to its value on the surface.

14. Consider two isolated spherical conductors each having net charge Q. The spheres have radii a and b, where $b > a$. Which sphere has the higher potential?

1. the sphere of radius a

2. the sphere of radius b

3. They have the same potential.

Answer: 1. The electrostatic potential for a conducting sphere varies inversely as the sphere radius.

DIELECTRICS & CAPACITORS

1. Consider a capacitor made of two parallel metallic plates separated by a distance d. The top plate has a surface charge density $+\sigma$, the bottom plate $-\sigma$. A slab of metal of thickness $l < d$ is inserted between the plates, not connected to either one. Upon insertion of the metal slab, the potential difference between the plates

1. increases.
2. decreases.
3. remains the same.

Answer: 2. Because the upper face of the slab becomes negatively charged and the lower face positively charged, they are attracted to the capacitor plates. We must do work on the system if we wish to remove the slab. Thus, the system has a lower potential energy when the slab is in place, and the potential difference between the plates is lower.

2. Consider two capacitors, each having plate separation *d*. In each case, a slab of metal of thickness *d*/3 is inserted between the plates. In case (*a*), the slab is not connected to either plate. In case (*b*), it is connected to the upper plate. The capacitance is higher for

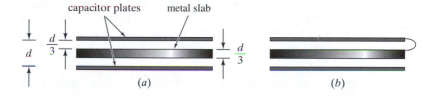

(*a*)   (*b*)

1. case (*a*).
2. case (*b*).
3. The two capacitances are equal.

Answer: 2. The system in case (*a*) is equivalent to two capacitors in series, each with plate separation *d*/3. The system in case (*b*) is equivalent to a single capacitor with plate separation *d*/3. Adding the capacitances in case (*a*), and using the fact that the capacitance varies inversely with the plate separation, we find that the capacitance is larger in case (*b*).

3. Consider a simple parallel-plate capacitor whose plates are given equal and opposite charges and are separated by a distance *d*. Suppose the plates are pulled apart until they are separated by a distance $D > d$. The electrostatic energy stored in the capacitor is

1. greater than
2. the same as
3. smaller than

before the plates were pulled apart.

Answer: 1. Since the oppositely charged plates are attracted to each other, work must be done to increase the separation between them. Thus, the electrostatic energy stored in the capacitor increases when the plates are separated.

4. A dielectric is inserted between the plates of a capacitor. The system is then charged and the dielectric is removed. The electrostatic energy stored in the capacitor is

   1. greater than

   2. the same as

   3. smaller than

   it would have been if the dielectric were left in place.

   Answer: 1. The dielectric is polarized when it is between the charged plates. The positively charged dielectric surface is attracted to the negatively charged plate, and the negatively charged dielectric surface is attracted to the positively charged plate. Thus, you must do work to remove the dielectric. As a result, the electrostatic energy stored in the capacitor increases when the dielectric is removed.

5. A parallel-plate capacitor is attached to a battery that maintains a constant potential difference V between the plates. While the battery is still connected, a glass slab is inserted so as to just fill the space between the plates. The stored energy

   1. increases.

   2. decreases.

   3. remains the same.

   Answer: 1. When the glass slab is between the plates, it becomes polarized, thereby decreasing the magnitude of the electric field. In order for the battery to maintain a constant potential difference across the plates, it must do work to deposit more charge on the plates. Thus, work is done on the system and the stored energy increases.

DC CIRCUITS

1. Consider two identical resistors wired in series (one behind the other). If there is an electric current through the combination, the current in the second resistor is

   1. equal to

   2. half

3. smaller than, but not necessarily half
the current through the first resistor.

Answer: 1. Charge is neither lost nor gained when the current passes
through the first resistor.

2. As more identical resistors R are added to the parallel circuit shown here,
the total resistance between points P and Q

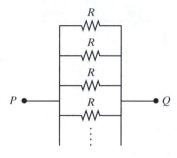

1. increases.
2. remains the same.
3. decreases.

Answer: 3. The potential difference is the same across each resistor, so
the same current flows through each resistor as would flow through an
isolated resistor of resistance R. The sum of the currents through all the
parallel resistors must equal the input current, so the total current increases as resistors are added and thus the total resistance decreases.

3. Charge flows through a light bulb. Suppose a wire is connected across the
bulb as shown. When the wire is connected,

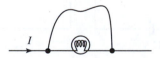

1. all the charge continues to flow through the bulb.
2. half the charge flows through the wire, the other half continues
through the bulb.
3. all the charge flows through the wire.
4. none of the above

Answer: 3. The wire has essentially zero resistance and is in parallel with the light bulb. Thus, all charge flows through the wire.

4. The circuit below consists of two identical light bulbs burning with equal brightness and a single 12 V battery. When the switch is closed, the brightness of bulb *A*

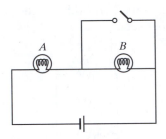

1. increases.
2. remains unchanged.
3. decreases.

Answer: 1. When the switch is closed, bulb *B* goes out because all of the current is through the wire parallel to the bulb. Thus, the total resistance of the circuit decreases, the current through bulb *A* increases, and it burns more brightly.

5. If the four light bulbs in the figure are identical, which circuit puts out more light?

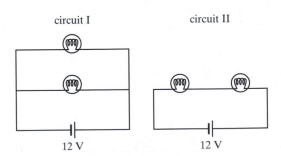

1. I.
2. The two emit the same amount of light.
3. II.

Answer: 1. The resistance of two light bulbs in parallel is smaller than that of two light bulbs in series. Thus, the current through the battery is greater for circuit I than for circuit II. Since the power dissipated is the product of current and potential difference, it follows that more power is dissipated in circuit I.

6. The light bulbs in the circuit are identical. When the switch is closed,

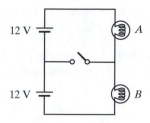

1. both go out.
2. the intensity of light bulb *A* increases.
3. the intensity of light bulb *A* decreases.
4. the intensity of light bulb *B* increases.
5. the intensity of light bulb *B* decreases.
6. some combination of 1–5 occurs.
7. nothing changes.

Answer: 7. The potential difference across the branch of the circuit containing the switch is zero. Thus, there is no current through it when the switch is closed, and nothing changes.

7. The light bulbs in the circuit are identical. When the switch is closed,

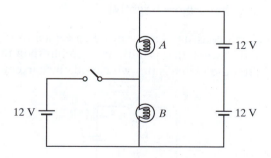

1. both go out.
2. the intensity of light bulb *A* increases.
3. the intensity of light bulb *A* decreases.

4. the intensity of light bulb *B* increases.

5. the intensity of light bulb *B* decreases.

6. some combination of 1–5 occurs.

7. nothing changes.

Answer: 7. Because the light bulbs are identical, the potential difference across each is 12 V, and so nothing happens with the switch is closed.

8. Two light bulbs *A* and *B* are connected in series to a constant voltage source. When a wire is connected across *B* as shown, bulb *A*

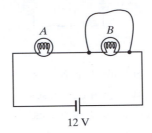

12 V

1. burns more brightly.

2. burns as brightly.

3. burns more dimly.

4. goes out.

Answer: 1. When the wire is connected across bulb *B*, all the current is through the wire. Bulb *B* is entirely bypassed because the wire has essentially zero resistance and is parallel to the bulb. The total resistance of the circuit decreases, the current increases, and thus the power dissipated increases. Bulb *A* burns more brightly.

9. A simple circuit consists of a resistor *R*, a capacitor *C* charged to a potential V_0, and a switch that is initially open but then thrown closed. Immediately after the switch is thrown closed, the current in the circuit is

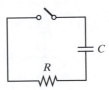

1. V_0/R.

2. zero.

3. need more information

Answer: 1. Immediately after the switch is thrown closed, the current is V_0/R. It decreases from this value to zero exponentially, with a time constant equal to RC.

10.   The three light bulbs in the circuit all have the same resistance. Given that brightness is proportional to power dissipated, the brightness of bulbs B and C together, compared with the brightness of bulb A, is

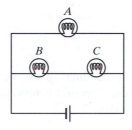

1. twice as much.
2. the same.
3. half as much.

Answer: 3. The potential difference across bulbs B and C in series is equal to the potential difference across bulb A. Since the power dissipated in a resistor of resistance R is V^2/R, where V is the potential difference across the resistor, the power dissipated by the series combination is one half the power dissipated by resistor (bulb) A.

11.   An ammeter A is connected between points a and b in the circuit below, in which the four resistors are identical. The current through the ammeter is

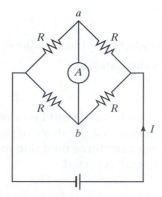

1. $I/2$.
2. $I/4$.
3. zero.
4. need more information

Answer: 3. Because the resistors are all identical, the potential difference between a and b is zero. Thus, there is no current through the ammeter.

MAGNETISM

1. On a computer chip, two conducting strips carry charge from P to Q and from R to S. If the current direction is reversed in both wires, the net magnetic force of strip 1 on strip 2

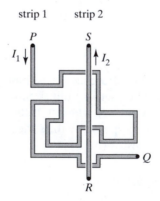

1. remains the same.
2. reverses.
3. changes in magnitude, but not in direction.
4. changes to some other direction.
5. other

Answer: 1. The magnetic force of strip 1 on strip 2 reverses when the current is reversed no matter what the shape of strip 1. If we also reverse the current strip 2, the magnetic force on it due to 1 is again what it was before the two currents were reversed.

2. A battery establishes a steady current around the circuit below. A compass needle is placed successively at points $P, Q,$ and R. The relative deflection of the needle, in descending order, is

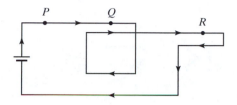

1. P, Q, R.
2. Q, R, P.
3. R, Q, P.
4. P, R, Q.
5. Q, P, R.

Answer: 5. Using the principle of superposition for the magnetic force, and adding only those forces due to wires in close proximity, we find that the compass needle has its largest deflection at Q and its smallest at R.

3. A charged particle accelerated to a velocity v enters the chamber of a mass spectrometer. The particle's velocity is perpendicular to the direction of the uniform magnetic field B in the chamber. After the particle enters the magnetic field, its path is a

1. parabola.
2. circle.
3. spiral.
4. straight line.

Answer: 2. The magnetic force on the particle is perpendicular to both the magnetic field and the particle's velocity. Thus, its path is circular.

4. Cosmic rays (atomic nuclei stripped bare of their electrons) would continuously bombard Earth's surface if most of them were not deflected by Earth's magnetic field. Given that Earth is, to an excellent approximation, a magnetic dipole, the intensity of cosmic rays bombarding its surface is greatest at the

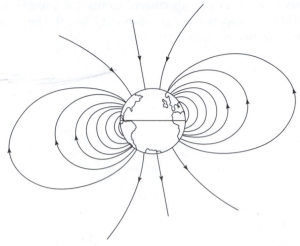

1. poles.
2. mid-latitudes.
3. equator.

Answer: 1. Assuming that the cosmic ray particles have velocities direct-ed radially toward Earth, the particles suffering the least deflection ap-proach Earth via its poles because it is here that the magnetic field lines most nearly point in the radial direction.

5. A $CuSO_4$ solution is placed in a container housing coaxial cylindrical cop-per electrodes. Electric and magnetic fields are set up as shown. Uncharged pollen grains added to the solution are carried along by the mobile ions in the liquid. Viewed from above, the pollen between the electrodes cir-culates clockwise. The pollen is carried by ions that are

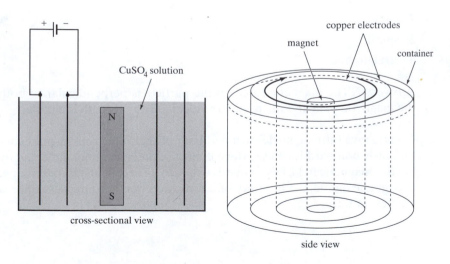

1. positive.
2. negative.
3. both positive and negative.
4. need more information

Answer: 3. Positive ions move radially inward due to the electric field between the cylinders. In the presence of the magnetic field, the right-hand rule indicates that they circulate clockwise. Negative ions move radially outward, but because of their negative charge, the right-hand rule gives a clockwise circulation for them too.

6. A sphere of radius R is placed near a long, straight wire that carries a steady current I. The magnetic field generated by the current is B. The total magnetic flux passing through the sphere is

1. $\mu_o I$.
2. $\mu_o I/(4\pi R^2)$.
3. $4\pi R^2 \mu_o I$.
4. zero.
5. need more information

Answer: 4. Because magnetic field lines close on themselves, the total magnetic flux through any closed surface is zero.

7. A rectangular loop is placed in a uniform magnetic field with the plane of the loop perpendicular to the direction of the field. If a current is made to flow through the loop in the sense shown by the arrows, the field exerts on the loop:

1. a net force.
2. a net torque.
3. a net force and a net torque.
4. neither a net force nor a net torque.

Answer: 4. According to the right-hand rule, all four sides of the loop are subject to forces that are directed outward, with the force on each side canceling that on the opposite side.

8. A rectangular loop is placed in a uniform magnetic field with the plane of the loop parallel to the direction of the field. If a current is made to flow through the loop in the sense shown by the arrows, the field exerts on the loop:

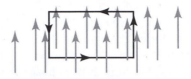

1. a net force.

2. a net torque.

3. a net force and a net torque.

4. neither a net force nor a net torque.

Answer: 2. According to the right-hand rule, the top side of the loop is subject to a force directed into the page and the bottom side to a force directed out of the page. These two forces exert a net torque on the loop (the effect of this torque is to direct the magnetic field of the loop parallel to the external field). The left and right side of the loop are not subject to forces.

9. When the switch is closed, the potential difference across R is

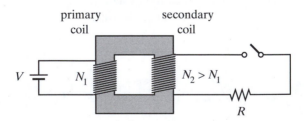

1. VN_2/N_1.

2. VN_1/N_2.

3. V.

4. zero.

5. insufficient information

Answer: 4. Because the input is connected to a constant voltage source no potential difference is induced in the secondary coil.

10. The primary coil of a transformer is connected to a battery, a resistor, and a switch. The secondary coil is connected to an ammeter. When the switch is thrown closed, the ammeter shows

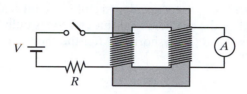

1. zero current.
2. a nonzero current for a short instant.
3. a steady current.

Answer: 2. When the switch is thrown closed, it momentarily produces a time-varying current in the primary coil of the transformer. The iron core then acts as an electromagnet and generates a transient current in the secondary coil via electromagnetic induction.

11. A long, straight wire carries a steady current I. A rectangular conducting loop lies in the same plane as the wire, with two sides parallel to the wire and two sides perpendicular. Suppose the loop is pushed toward the wire as shown. Given the direction of I, the induced current in the loop is

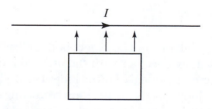

1. clockwise.
2. counterclockwise.
3. need more information

Answer: 2. As the loop approaches the wire, the magnetic flux through the loop will increase. To counteract this increase in magnetic flux, a current will start flowing to create a counteracting magnetic flux through the loop. Because the magnetic field of the wire points into the page at the location of the loop, the counteracting magnetic field of the loop must point out of the page. This requires a counterclockwise current.

12. In figure (a), a solenoid produces a magnetic field whose strength increases into the plane of the page. An induced emf is established in a conducting loop surrounding the solenoid, and this emf lights bulbs *A* and *B*. In figure (b), points *P* and *Q* are shorted. After the short is inserted,

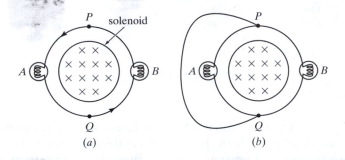

(a)   (b)

1. bulb *A* goes out; bulb *B* gets brighter.
2. bulb *B* goes out; bulb *A* gets brighter.
3. bulb *A* goes out; bulb *B* gets dimmer.
4. bulb *B* goes out; bulb *A* gets dimmer.
5. both bulbs go out.
6. none of the above

Answer: 1. Although one might expect both bulbs to go out when shorted, only bulb *A* does. Together with the original loop, the extra wire between *P* and *Q* creates a total of three loops (see following figure): (*i*) the original circular loop; (*ii*) the heavy loop consisting of the extra wire and the right half of the original loop; and (*iii*) the crescent-shaped conducting loop consisting of the extra wire plus the dashed wire connected to bulb *A*. Because there is no changing flux through loop (*iii*), bulb *A* will not light. Loop (*ii*), however, encloses a changing magnetic flux and thus bulb *B* lights. Loop (*i*) also encloses a changing magnetic flux, but it is "easier" for the electrons to flow around loop (*ii*) because the extra wire has zero resistance, whereas the dashed wire containing bulb *A* has finite resistance. Since the net resistance in the circuit is now less than it was before the wire was added, bulb *B* gets brighter.

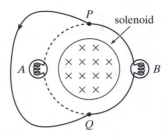

13.   When the switch is closed, the current through the circuit exponentially approaches a value $I=\mathscr{E}/R$. If we repeat this experiment with an inductor having twice the number of turns per unit length, the time it takes for the current to reach a value of $I/2$

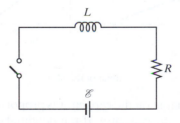

1. increases.

2. decreases.

3. is the same.

Answer: 1. As the number of turns is increased the inductor generates a larger magnetic field for a given current. Therefore a change in current will generate a larger change in flux and hence the back-emf goes up as the number of turns is increased. As a result, the current will grow more slowly than before. Mathematically, the time constant for the exponential increase of the current is L/R. As the number of turns per unit length increases, L increases and the time constant increases. Thus, it will take longer for the current to reach a value of $I/2$.

AC CIRCUITS

1. A capacitor is connected to a varying source of emf. Given the behavior of $\mathscr{E}$ shown, the current through the wires changes according to:

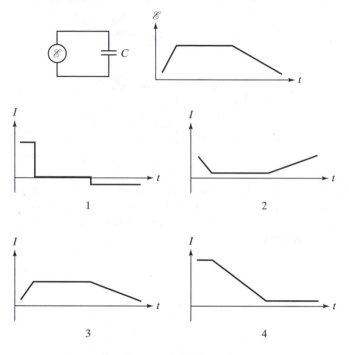

5. none of the above

Answer: 1. The charge on the capacitor is proportional to the potential difference applied to the capacitor. If the potential difference increases linearly, then the charge will increase linearly, and so there must be a steady current (charge per unit time) through the wires. Similarly, if the potential difference is constant, the charge is constant and hence there is zero current. Finally, when the potential difference decreases, the charge decreases too and so charge must flow back in the opposite direction. Mathematically, the current is proportional to the time derivative of the emf.

2. A capacitor is connected to a varying source of emf. The work done by the source during the time intervals $a, b,$ and c is

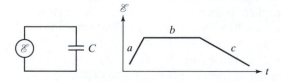

1. positive, negative, and zero, respectively.
2. negative, positive, and zero, respectively.
3. always positive.
4. positive, zero, and negative, respectively.
5. always negative.
6. zero, positive, and zero, respectively.
7. zero, negative, and zero, respectively.

Answer: 4. During time interval a, as the emf is increasing, the charge on the capacitor is increasing and hence the amount of energy stored in the capacitor increases. To store energy, positive work must be done. During b, when the emf is constant, no work is done as the energy stored in the capacitor is constant. During c, the charge on the capacitor decreases and hence energy flows back out of the capacitor to the source.

3. The phasor diagrams below represent three oscillating emfs having different amplitudes and frequencies at a certain instant of time $t = 0$. As t increases, each phasor rotates counterclockwise and completely determines a sinusoidal oscillation. At the instant of time shown, the magnitude of $\mathscr{E}$ associated with each phasor given in ascending order by diagrams

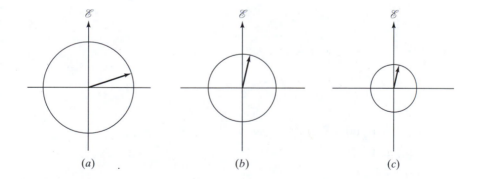

(a)   (b)   (c)

1. (a), (b), and (c).
2. (a), (c), and (b).
3. (b), (c), and (a).
4. (c), (a), and (b).
5. none of the above

6. need more information

Answer: 3. The instantaneous value of $\mathscr{E}$ is given by the projection of the phasor onto the vertical axis.

4. Consider the pairs of phasors below, each shown at $t = 0$. All are characterized by a common frequency of oscillation ω. If we add the oscillations, the maximum amplitude is achieved for pair

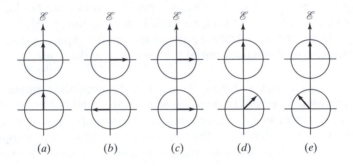

(a) (b) (c) (d) (e)

1. (a).
2. (b).
3. (c).
4. (d).
5. (e).
6. (a), (b), and (c).
7. (a) and (c).
8. (b) and (c).
9. need more information

Answer: 7. Pairs (a) and (c) are in phase, so the amplitude of their sum equals the sum of their individual amplitudes.

5. Consider the oscillating emf shown below. Which of the phasor diagrams correspond(s) to this oscillation:

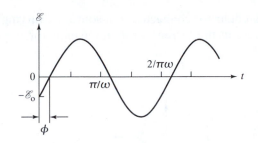

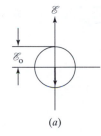

(a)

(b)

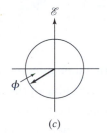

(c)

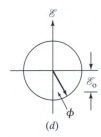

(d)

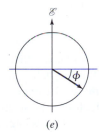

(e)

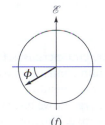

(f)

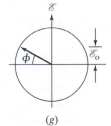

(g)

1. all but (b) and (c)
2. all
3. (e), (f), and (g)
4. (d)
5. (e)
6. all but (a) and (d)
7. (d) and (e)
8. none

Answer: 5. The phasor, rotating counterclockwise, determines the sinusoidal oscillation. Only the phasor in (e) gives an initial value of $-\mathscr{E}_0$ and an initial phase of $-\phi$.

6. Consider an inductor connected to a source of varying emf. If the graph below represents the current through the inductor, the work done by the source is:

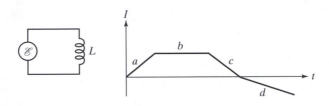

1. positive during time intervals a, b, and c.
2. zero during b, positive during d.
3. positive during a, negative during d.
4. positive during b, negative during d.
5. positive during a, zero during b, negative during c and d.
6. none of the above

Answer: 2. During a, as the emf increases, the magnetic field inside the inductor grows and so the source must do positive work. During b, the magnetic field in the conductor is constant and no work is done. During c, the field decreases and thus energy flows back from the inductor to the source and hence the work is negative. During d, however, the field grows again (but in the opposite direction) and hence positive work must be done. Mathematically, the work done by the source is proportional to the current times its time rate of change.

7. The light bulb has a resistance R, and the emf drives the circuit with a frequency ω. The light bulb glows most brightly at

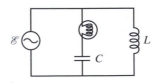

1. very low frequencies.
2. very high frequencies.
3. the frequency $\omega = 1/\sqrt{LC}$.

Answer: 2. At very high frequencies, the capacitor has essentially zero impedance and the inductor essentially infinite impedance and so the current through the light bulb is largest.

8. For the *RLC* series circuit shown, which of these statements is/are true:

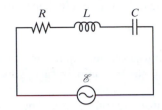

(*i*)   Potential energy oscillates between *C* and *L*.

(*ii*) The source does no net work: Energy lost in *R* is compensated by energy stored in *C* and *L*.

(*iii*) The current through *C* is 90° out of phase with the one through *L*.

(*iv*) The current through *C* is 180° out of phase with the one through *L*.

(*v*)   All energy is dissipated in *R*.

1. all of them
2. none of them
3. (*v*)
4. (*ii*)
5. (*i*), (*iv*), and (*v*)
6. (*i*) and (*v*)
7. none of the above

Answer: 6. Only statements (*i*) and (*v*) are correct. Because the capacitor and the inductor are not sources of energy, energy dissipated by the resistor cannot be compensated for, so (*ii*) is incorrect. Furthermore, since the current is in phase throughout the circuit (*iii*) and (*iv*) are incorrect.

ELECTRODYNAMICS

1. As the capacitor shown below is charged with a constant current *I*, at point *P* there is a

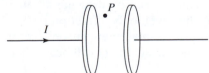

1. constant electric field.

2. changing electric field.

3. constant magnetic field.

4. changing magnetic field.

5. changing electric field and a magnetic field.

6. changing magnetic field and an electric field.

7. none of the above.

Answer: 5. As the charge on the capacitor increases, the electric field increases. The changing electric field, in turn, generates a magnetic field.

2. For a charging capacitor, the total displacement current between the plates is equal to the total conduction current I in the wires. The capacitors in the diagram have circular plates of radius R. In (a), points A and B are each a distance $d > R$ away from the line through the centers of the plates; in this case the magnetic field at A due to the conduction current is the same as that at B to the displacement current. In (b), points P and Q are each a distance $r < R$ away from the center line. Compared with the magnetic field at P, that at Q is

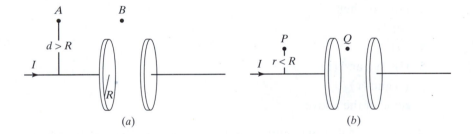

1. bigger.

2. smaller.

3. the same.

4. need more information.

Answer: 2. The magnetic field at P is proportional to the entire conduction current, as can be seen by drawing an amperian loop of radius r through P. At Q, however, only a fraction r^2/R^2 of the changing electric flux, and hence of the displacement current, contributes to the magnetic field. Thus, the magnetic field at Q will be smaller than that at A.

3. A planar electromagnetic wave is propagating through space. Its electric field vector is given by $\mathbf{E} = E_0 \cos(kz - \omega t)\hat{\mathbf{x}}$ Its magnetic field vector is

1. $\mathbf{B} = B_0 \cos(kz - \omega t)\hat{\mathbf{y}}$
2. $\mathbf{B} = B_0 \cos(ky - \omega t)\hat{\mathbf{z}}$
3. $\mathbf{B} = B_0 \cos(ky - \omega t)\hat{\mathbf{x}}$
4. $\mathbf{B} = B_0 \cos(kz - \omega t)\hat{\mathbf{z}}$

Answer: 1. From the expression for the electric field vector we see that the wave is propagating in the z direction with the electric field aligned along the x axis. Propagating electromagnetic waves have orthogonal electric and magnetic fields.

4. At a fixed point, P, the electric and magnetic field vectors in an electromagnetic wave oscillate at angular frequency ω. At what angular frequency does the Poynting vector oscillate at that point?

   1. 2ω

   2. ω

   3. $\omega/2$

   4. 4ω

Answer: 1. The Poynting vector is proportional to the cross product of the electric and magnetic field vectors. Since both fields oscillate sinusoidally with frequency ω, trigonometric identities show that their product is a sinusoidal function of frequency 2ω.

5. Which gives the largest average energy density at the distance specified and thus, at least qualitatively, the best illumination

   1. a 50-W source at a distance R.

   2. a 100-W source at a distance $2R$.

   3. a 200-W source at a distance $4R$.

Answer: 1. The energy density at the specified point is proportional to the power divided by the surface area of the indicated sphere.

MODERN PHYSICS

1. The best color to paint a radiator, as far as heating efficiency is concerned, is

   1. black.

   2. white.

   3. metallic.

   4. some other color.

   5. It doesn't really matter.

Answer: 1. The best absorber is also the best radiator.

2. A beam of ultraviolet light is incident on the metal ball of an electroscope. Which statement(s) is/are true?
   1. If the electroscope was initially positively charged, it discharges.
   2. If the electroscope was initially negatively charged, it discharges.
   3. Both of the above.
   4. Neither of the above.

Answer: 2. When the ultraviolet radiation strikes the electroscope, photoelectrons are emitted from the metal. Thus, an electroscope which is initially negatively charged will discharge. If the electroscope is initially positively charged, any electron that is emitted is immediately attracted back to the electroscope. The same happens if the electroscope is initially neutral because each emitted photoelectron leaves a positive charge behind in the electroscope.

3. A beam of ultraviolet light is incident on the metal ball of an electroscope that is initially uncharged. Does the electroscope acquire a charge?
   1. Yes, it acquires a positive charge.
   2. Yes, it acquires a negative charge.
   3. No, it does not acquire a charge.

Answer: 3. Each electron that leaves the electroscope via photoemission is immediately attracted back to the electroscope because of electrostatic attraction.

4. A xenon arc lamp is covered with an interference filter that only transmits light of 400-nm wavelength. When the transmitted light strikes a metal surface, a stream of electrons emerges from the metal. If the intensity of the light striking the surface is doubled,
   1. more electrons are emitted in a given time interval.
   2. the electrons that are emitted are more energetic.
   3. both of the above.
   4. neither of the above.

Answer: 1. The increased intensity of the light source will increase the number of photoelectrons emitted per unit time. Since the frequency, and hence the photon energy, of the additional light source is equal to that of the original source, the energy of the ejected electrons will not change.

5. A xenon arc lamp is covered with an interference filter that only transmits light of 400-nm wavelength. When the transmitted light strikes a metal surface, a stream of electrons emerges from the metal. The interference filter is then replaced with one transmitting at 300 nm and the lamp adjusted so that the intensity of the light striking the surface is the same as it was for the 400-nm light. With the 300-nm light,

1. more electrons are emitted in a given time interval.

2. the electrons which are emitted are more energetic.

3. both are true.

4. both are false.

Answer: 2. The shorter wavelength of the light means that the photons are more energetic. Consequently the ejected electrons have more energy.

6. In a Michelson interferometer, a beam of light is split into two parts of equal intensity, and the two parts are subsequently recombined to interfere with one another. When a single photon is sent through the interferometer, the photographic plate shows

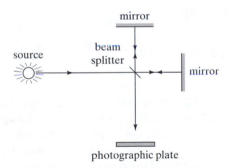

1. a single dot somewhere on the plate because the photon chooses one of the two paths through the splitter and then returns and strikes the plate.

2. a single dot, which is more likely to lie in some regions than others, because of the interference between the two paths.

3. an interference pattern because the interferometer splits the photon into two waves that subsequently interfere at the plate.

Answer: 2. Single photons coming through the splitter hit the plate at random, producing a scatter of individual points. Once enough photons have hit the plate, however, an interference pattern emerges. The photons register on the plate as single points, but the likelihood of a photon hitting a particular point is governed by the interference pattern.

7. Single photons are directed, one by one, toward a double slit. The distribution pattern of impacts that make it through to a detector behind the slits is identical to an interference pattern. We now repeat this experiment, but block slit 1 for the first half of the experiment and slit 2 for the second half. The distribution of impacts in the second experiment is

    1. the same as in the first experiment.
    2. the sum of the distributions one gets for each slit separately.
    3. neither of the above.

    Answer: 2. When one slit is covered, the pattern of photon impacts produces a distribution characteristic of single-slit interference. When the second slit is covered, the photon impacts again distribute themselves into a single-slit pattern. The final result is the sum of the two.

8. Thompson observed that cathode beams can pass undeflected through crossed electric and magnetic fields. Which of the following quantities must then be common to the particles making up these beams

    1. mass
    2. size
    3. magnitude of charge
    4. sign of charge
    5. sign and magnitude of charge
    6. velocity

    Answer: 6. If the particles are not deflected, the net force on the particles must be zero and thus the electric and magnetic forces are equal and opposite. The electric force and the magnetic force are both independent of the mass and size of the particles. They are, however, proportional to the charge q of the particles, and the magnetic force is also proportional to the velocity v of the particles. Changing the mass, size, or charge of the particles therefore does not affect the balance of the electric and magnetic forces. Changing the velocity, in contrast changes the magnetic force, but not the electric force.

9. A cathode beam passes undeflected through crossed electric and magnetic fields. When the electric field is switched off, the beam splits up in several beams. This splitting is due to the particles in the beam having different

    A. masses.
    B. velocities.
    C. charges.
    D. none of the above

Answer: A and C. The deflection of the particles is proportional to the force on them divided by their mass. When the electric field is switched off, the only force on the particles is the magnetic force, which depends on the particles' charge and velocity. Because the particles go undeflected through the crossed field, we know they all have the same velocity, and so the deflection depends on the mass and the charge of the particles.

10. Cathode rays are beams of electrons, but the electrons are not deflected downward by gravity because

1. the effect of gravity on electrons is negligible.

2. the electrons go so fast there's no time to fall.

3. of air resistance.

4. the electrons are quantum particles and not classical particles.

5. the electric charge prevents electrons from feeling gravity.

6. other

Answer: 2. The electrons travel so rapidly in the horizontal direction that, during their horizontal flight, the distance they fall vertically due to gravity is negligible.

11. An emission spectrum for hydrogen can be obtained by analyzing the light from hydrogen gas that has been heated to very high temperatures (the heating populates many of the excited states of hydrogen). An absorption spectrum can be obtained by passing light from a broadband incandescent source through hydrogen gas. If the absorption spectrum is obtained at room temperature, when all atoms are in the ground state, the absorption spectrum will

1. be identical to the emission spectrum.

2. contain some, but not all, of the lines appearing in the emission spectrum.

3. contain all the lines seen in the emission spectrum, plus additional lines.

4. look nothing like the emission spectrum.

Answer: 2. When radiation passes through the room-temperature gas, electrons are excited from the ground state to excited states; each line in the absorption spectrum represents one of these ground → excited state transitions. When the hydrogen atoms are heated to high temperatures, transitions from excited states to the ground state occur, but there are additional transitions from higher excited states to lower ones that are above the ground state. Therefore the emission spectrum contains all the excited → ground state counterparts of the absorption spectrum plus more lines representing high-excited → low-excited state transitions.

12

CONCEPTUAL EXAM QUESTIONS

On the following pages are conceptual questions of the type I have used on my examinations since 1991. Typically these conceptual questions make up half the examination, with the other half devoted to more traditional computational problems. Both types contribute the same total number of points to the grade.

Many of these conceptual questions are essay questions, and the most convenient grading scheme for them is one that mirrors the review practice of professional scientific journals: Each problem is worth a maximum of three points, and the score is arrived at via the following guidelines:

Review rating	Points	Rating of answer
publish as is	3	perfect or nearly perfect
publish after minor revision	2	small errors that require revision
needs major revision	1	substantial errors
reject	0	little or no relevance to question

At first it may seem that this scheme will result in a loss of grading accuracy, but even if an exam contains only five problems, the overall grading accuracy is better than 7% (which I believe is better than practically feasible). Most of my exams contain at least seven problems, and so the situation is even better. The major advantages of the rating scheme are that it leads to a greater grading consistency and to fewer students arguing about their grade —the difference between a 1 and a 2, for instance, is much more clear-cut than that between a 6 and a 7 on a ten-point scale. When a conceptual question contains multiple parts, the scheme is applied to the question *as a whole* rather than the individual parts. After correcting each part, noting errors, and classifying them as minor or serious, I ask myself the question, "Would I accept this paper for publication?" and, based on the answer, I give an overall score for the problem.

KINEMATICS

1. Two stones are released from rest at a certain height, one after the other. (*a*) Will the difference in their speeds increase, decrease, or stay the same? (*b*) Will their separation distance increase, decrease, or stay the same? (*c*) Will the time interval between the instants at which they hit the ground be smaller than, equal to, or larger than the time interval between the instants of their release?

2. As a space shuttle burns up its fuel after take-off, it gets lighter and lighter and its acceleration larger and larger. Between the moment it takes off and the time at which it has consumed nearly all of its fuel, is its *average speed* larger than, equal to, or smaller than half its *final speed?*

3. You are in an elevator that is rising at constant velocity. Suddenly you drop your keys; it so happens that when they strike the floor they are as high above ground level as when they left your hand. The keys fall dead on the floor without bouncing. Make a single graph showing qualitatively the height above ground of both the keys and the elevator as a function of time, starting from before the keys are released until after they strike the floor.

4. (*a*) Can astronauts floating in orbit tell which objects within their spaceship would be heavy or light on Earth even though everything in the ship is effectively weightless? Explain. (*b*) Could an astronaut use a "weightless" hammer in space to drive a nail through a piece of wood? Explain.

5. A battleship simultaneously fires two shells toward two enemy ships, one close by (*A*), one far away (*B*). The shells leave the battleship at different angles and travel along the parabolic trajectories indicated below. Which of the two enemy ships gets hit first? Explain.

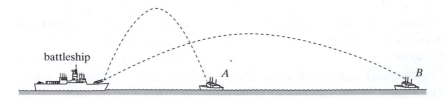

6. Cite an example of a body that accelerates while traveling at constant speed. Is it possible for a body to be accelerating while traveling at constant velocity? Explain.

NEWTON'S LAWS

1. You are riding your bike along a flat country road. Indicate the direction and relative magnitude of the frictional forces on the front and rear tires

in the following situations: (*a*) you are accelerating; (*b*) you are pedaling along at a steady pace; (*c*) you are braking. Both the brake and the pedals work on the rear wheel; there is no brake on the front wheel.

2.   Identify all the action-reaction pairs that exist for a horse pulling a cart. Include Earth in your examination, but ignore air resistance. Make sure your notation makes clear which force acts on which object.

3.   What is the smallest force with which you can raise an object on Earth into the air?

4.   (*a*) A car is stationary on a flat parking lot. The force of gravity acts downward, and an equal and opposite normal force acts upward. State the law according to which these forces are equal and opposite. (*b*) When the car accelerates forward, which force is responsible for this acceleration? State clearly which body exerts this force, and on which body the force acts.

5.   A horse is pulling a wagon at constant velocity over a flat, horizontal road. Draw a free-body diagram of all the forces acting on the horse and another one of all the forces acting on the wagon. In your drawing, indicate the relative magnitude of the forces and identify any third-law force pairs. Also identify forces of equal magnitude that are not third-law pairs.

6.   A basketball player is jumping vertically upward in order to land a shot. Her legs are flexed and pushing on the floor so that her body is accelerated upward. (*a*) Draw free-body diagrams of the player's body and Earth. Show the relative magnitudes of the various forces and describe each in words (*i.e.*, contact, gravitational, etc., and indicate which object exerts that force and on what). Identify the action-reaction pairs. (*b*) Repeat this exercise for the situation immediately after the player's body breaks contact with the floor. (*c*) Finally, consider, in the same manner, the situation at the top of the jump.

7.   A toy train travels around a loop-the-loop track, as shown below. (*a*) Is there a normal force exerted by the track on the train at the instant the train is at the top of the loop? (*b*) Why is it that riders feel weightless at the top of certain roller coasters?

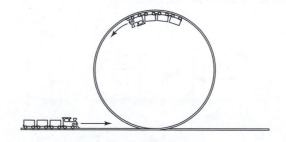

8. A monkey clings to a rope that passes over a pulley. The monkey's weight is balanced by the mass m of a block hanging at the other end of the rope; both monkey and block are motionless. In order to get to the block, the monkey climbs a distance L (measured along the rope) up the rope. (*a*) Does the block move as a result of the monkey's climbing? (*b*) If so, in which direction and by how much?

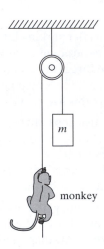

9. A child sits on a pedestal that rests on the platform of a merry-go-round. He holds a bob suspended on a string. Draw free-body diagrams for the child, the bob, the string, the pedestal, and the platform. Indicate the magnitudes of the forces by the relative lengths of the arrows you draw. Describe each force and indicate the third-law pairs.

10. A weightlifter and a barbell are both at rest on a large scale. The weightlifter begins to lift the barbell, ultimately holding it motionless above her head. Does the scale reading ever differ from the combined weight of the two bodies at any time during the lift? Explain.

11. Two blocks sitting on a frictionless table are pushed from the left by a horizontal force, as shown below.

(*a*) Draw a free-body diagram for each of the blocks. (*b*) Express, in terms of the quantities given in the figure, the force of contact between the two blocks. (*c*) What is the acceleration of the blocks?

Suppose now that, instead of a force on the left block, a force of equal magnitude but opposite direction is applied to the block on the right.

(*d*) What is the force of contact between the two blocks in this case? (*e*) What is the acceleration of the blocks? (*f*) Do your answers to parts *b* through *e* make sense? Verify.

12. Two teams, *A* and *B*, are competing in a tug-of-war. Team *A* is stronger, but neither team is moving because of friction. Draw free-body diagrams for team *A*, for team *B*, and for the rope.

13. The spring in configuration (*a*) is stretched 0.10 m. How much will the same spring be stretched in configuration (*b*)?

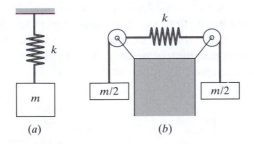

14. In the situation below, a person pulls a string attached to block *A*, which is in turn attached to another, heavier block *B* via a second string. (*a*) Which block has the larger acceleration? (*b*) How does the force of string 1 on block *A* compare with the force of string 2 on block *B*? Explain your answers.

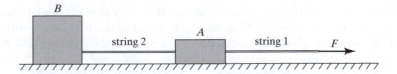

15. For the swinging pendulum shown below, (*a*) make a diagram indicating the acceleration of the bob at positions *P* (the end of the swing) and *Q* (the bottom of the swing), and (*b*) draw force diagrams of the bob when it is at positions *P* and *Q*. In each diagram, show the net (resultant) force acting on the bob, if any.

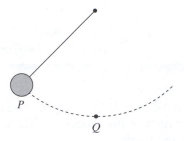

16. A heavy pendulum bob is swinging back and forth when the string supporting it suddenly breaks. Ignoring the mass of the string and air friction, describe the subsequent motion of the bob if the break occurs when the bob is (*a*) at its lowest point and (*b*) at its highest point.

17. A block is placed on a planar sheet that is pivoted at one end. The free side of the sheet is then raised very slowly, as shown below. When the sheet is first raised, friction between block and sheet keeps the block from moving. At a certain angle, however, the block begins to slide down the inclined sheet. (*a*) If the sheet is kept at this angle, will the acceleration of the block be zero, constant, or neither? Why? (*b*) If the condition $\mu_k < \mu_s$ were not true, what would happen to the block?

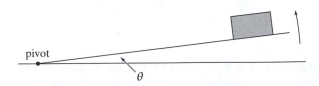

WORK AND ENERGY

1. Two people push in opposite directions on a block that sits atop a frictionless surface (The soles of their shoes are glued to the frictionless surface). If the block, originally at rest at point *P*, moves to the right without rotating and ends up at rest at point *Q*, describe qualitatively how much work is done on the block by person 1 relative to that done by person 2?

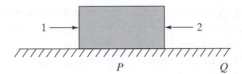

2. (*a*) When an object's kinetic energy is increasing, must its potential energy be decreasing? Explain. (*b*) If a rocket engine delivers a constant thrust (force on the rocket), does it deliver more power as the rocket speeds up? Explain.

3. A marble is dropped from the top of a skyscraper. Is the work done by Earth's gravitational force on the marble equal to, smaller than, or larger than that done by the marble's gravitational force on Earth? Explain.

4. Given that the gravitational force on a body is proportional to the body's mass, why doesn't a heavy body fall more rapidly than a light body?

5. It is possible to have a non-zero net force on an object that does no work on the object? Give an example of this for (*a*) an object going in a straight line and (*b*) an object along a curved trajectory. Explain.

6. Suppose you're on a rooftop and you throw a ball downward to the ground below with an initial speed *v*. You then throw an identical ball upward with the same initial speed *v*. After rising, this ball also falls to the ground. At the moment each strikes the ground, how do the speeds of the balls compare? Use conservation of mechanical energy in your reasoning.

MOMENTUM & COLLISIONS

1. Identical constant forces push two identical masses *A* and *B* continuously from a starting line to a finish line. If *A* is initially at rest and *B* is initially moving to the right, which mass has the larger change in momentum?

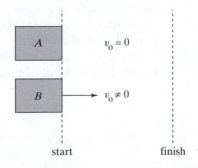

2. A moving object collides with an object initially at rest. (*a*) Is it possible for both objects to be at rest after the collision? (*b*) Can one of them be at rest after the collision? (*c*) Is it possible to have a collision in which all kinetic energy is lost? Explain each answer, and if you answer "yes" to any part, cite an example.

3. You are to knock over a gigantic bowling pin at a fair by throwing a ball at it. You have a choice between two balls, one that collides elastically and another that collides inelastically. Which do you choose and why?

4. Suppose you are in an automobile accident. (*a*) All other things being equal, are you better off in a more massive car or a lighter one? Why? (*b*) Is it better to be in a car that crumples on impact rather than one that holds together stiffly? Why?

5. A very elastic superball is dropped from a height *h* onto the ground. The ball rebounds at very nearly its original speed. (*a*) Is the ball's momentum conserved during any part of this process? (*b*) If we consider the ball and Earth as our system, during what parts of the process is momentum conserved? (*c*) Answer (*a*) and (*b*) for the case of a piece of putty that falls and sticks to the ground.

6. Suppose a golf ball is hurled at a heavy bowling ball initially at rest and bounces elastically from the bowling ball. After the collision, which ball has the greater momentum? Which has the greater kinetic energy? Why?

ROTATIONS

1. Suppose you are standing on the center of a merry-go-round that is at rest. You are holding a rotating bicycle wheel over your head so that its rotation axis is pointing upward. The wheel is rotating counterclockwise when observed from above along your axis of rotation. (*a*) Suppose you now move the wheel so that its axis is horizontal. What happens to you? (*b*) What happens if you then point the axis of the wheel downward so that the wheel rotates clockwise as viewed from above?

2. The figure below shows the paths taken by several particles. In (*a*)–(*d*), the particles move with constant speed, and in (*e*), the motion is back and forth. With respect to the origins *O*, state the direction of each angular momentum vector (or indicate if the angular momentum is zero). In each case, is the angular momentum independent of time?

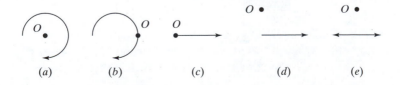

(*a*)   (*b*)   (*c*)   (*d*)   (*e*)

3. A ball initially at rest rolls without slipping down an inclined plane, as shown below. (*a*) Make a diagram of the ball on the incline showing all forces acting on the ball. Describe each force in words. (*b*) Which force described in part (*a*) causes the ball to roll by creating a torque about its center?

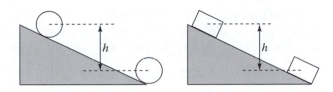

Now consider a block sliding down an identical inclined plane. The block travels the same vertical distance as the ball before arriving at the bottom. (*c*) Which arrives at the bottom with more total kinetic energy? Why? (*d*) Which arrives at the bottom with more linear momentum? Why?

OSCILLATIONS, WAVES, AND SOUND

1. Suppose you are at rest with respect to a source of sound and a strong wind is blowing in your direction. Describe the effect, if any, of the wind on (*a*) the observed frequency, (*b*) the observed wavelength, and (*c*) the wave speed?

2. Is it possible to move a source of sound relative to a stationary observer so that there is no shift in the sound frequency? If so, give an example.

3. A spring hanging vertically with a mass m suspended from it elongates an amount Δl. If pulled a little farther and then let go, the mass oscillates with a measured period, T. Discuss how you could use this arrangement to determine the value of g.

FLUIDS

1. Consider two identical ships, one loaded with a cargo of cork (which floats) and the other empty. Which ship displaces more water?

2. A piece of cork floats in a pail of water that rests on the floor of an elevator. If the elevator accelerates upward, how does the depth at which the cork floats change?

3. Use the principle behind Bernoulli's equation to explain the following phenomena: (*a*) smoke rises in a chimney faster when a breeze is blowing and (*b*) when a fast-moving train passes a train at rest, the two tend to be drawn together.

ELECTROSTATICS

1. In the figure below, two uncharged conductors of identical mass and shape are suspended from a ceiling by nonconducting strings. The conductors are given charges $q_1=Q$ and $q_2=3Q$. (a) After charging, which of the angles θ_1 and θ_2 that the two strings make with the vertical is larger, or are they equal?

    The two conductors are now brought together and made to touch. (b) Which of the two new angles θ_1' and θ_2' is the larger, or are they equal? (c) How do the angles θ_1' and θ_2' compare with the old angles θ_1 and θ_1?

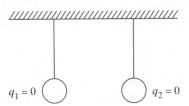

2. In the following figure, the dashed line denotes a Gaussian surface enclosing part of a distribution of four positive charges. (a) Which charges contribute to the electric field at P? (b) Is the value of the flux through the surface, calculated using only the electric field due to q_1 and q_2, greater than, equal to, or less than that obtained using the field due to all four charges?

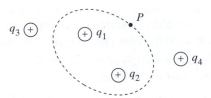

3. The following figure depicts three charges and a Gaussian surface. (a) Which charges contribute to the net flux through the Gaussian surface? (b) Which of the charges contribute to the field at a given point on the surface? (c) Compare your answers to (a) and (b) and explain why they are the same or different.

    Suppose the net charge enclosed by a surface is zero. (d) Does it follow that the field is zero at all points on the surface? (e) Is the reverse true (i.e., if the field is zero at all points on the surface, is the net charge enclosed zero)?

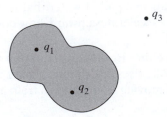

4. Suppose two charged conducting spheres of different radii r_1 and $r_2 > r_1$ are connected by a conducting wire. Which sphere has the greater charge density?

5. If the electric field E equals zero at a given point, must the potential difference V also equal zero at that point? Give an example to prove your answer.

6. A spherical rubber balloon has a charge uniformly distributed over its surface. As the balloon is inflated, how does the electric field E vary (a) outside the balloon, at some point well away from the surface? (b) at the outer surface of the balloon? (c) inside the balloon? Assume the balloon remains spherical during inflation.

7. (a) In the figure below, a point charge $+q$ is placed midway between two identical point charges $+Q$. Is q in equilibrium? If so, is the equilibrium stable or unstable? (b) Answer the same two questions for the case where $+q$ is changed to $-q$ but $+Q$ remains positive.

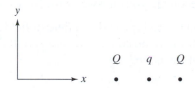

8. A charged rod is placed near an uncharged, nonconducting sphere suspended from a nonconducting wire, as shown below. Neither the rod nor anything else touches the sphere. (a) Will the rod and sphere exert forces on one another? If so, make a sketch showing the direction of the forces. (b) Would your answer to (a) change if the sphere were made out of conducting material? (c) What is the total charge on the sphere after the rod is placed close to it? (d) Do your answers to (a) and (c) violate Coulomb's law? Explain.

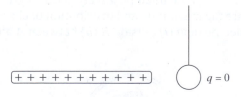

9. Rod A is a positively charged insulator. Bob C and a second rod B are in contact with each other and made from conducting material. Rod B is fixed, and C is suspended from a wire and free to swing. Briefly describe what happens when A is brought near B.

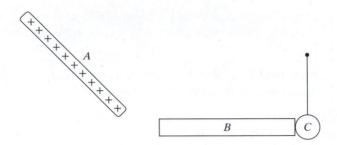

10. Two identical charges $+q$ are a distance L apart. Is it possible to place a third charge q' somewhere, such that none of the charges experience any force? If so, where and what is q'? If not, why not?

11. A point charge Q is at the center of a spherical conducting shell. There is a point charge q outside the shell. (a) Does q experience a force? (b) Does Q experience a force? (c) If there is a difference in the forces experienced by the charges, reconcile your answer with Newton's third law.

12. A hollow insulating cone is placed in a field of field strength E. What is the ratio of the flux through the conical surface A to the flux through the open cross-section B of the cone?

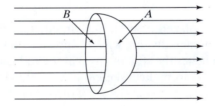

13. In the figure below, A is a solid metallic sphere and B a concentric metallic shell. Suppose A is charged positively and B is grounded. (a) Qualitatively compare the magnitude and distribution of charges on A and B. (b) Is there an electric field (i) outside B (ii) between A and B; (iii) inside A? Explain.

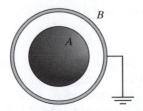

14. Suppose a positive point charge q is located at the center of a spherical metal shell. What charges appear on (a) the outer surface and (b) the inner surface of the sphere? (c) Suppose you bring an (uncharged) metal object near the sphere. Will your answers to parts (a) or (b) above change? Will the way charge is distributed over the sphere change?

15. Consider the electric field inside a charged rubber (insulating) balloon. Is E necessarily zero inside the balloon if it is (a) spherical, or (b) sausage-shaped? Assume the charge to be distributed uniformly over the surface for each shape. (c) How would the situation change, if at all, if the balloon were coated with a thick layer of conducting paint on its outside surface?

16. A positive charge moves in the direction of a uniform electric field. Does its potential energy increase or decrease? Does the electric potential increase or decrease?

17. If the electric potential at some point is zero, does it follow that there are no charges in the vicinity of that point? Explain.

DC CIRCUITS

1. Suppose a resistor and a capacitor are connected in series to a battery. Does the value of the resistor affect how much charge the capacitor stores? If yes, explain how. If no, what is the effect of the resistor?

2. Two light bulbs both operate from 110 V, but one has a power rating of 40 W whereas the other has a rating of 75 W. (a) Which bulb has the higher resistance? (b) Which carries the greater current?

3. The following circuit has three identical light bulbs connected to an ideal battery. (a) How do the brightnesses of the three bulbs compare? (b) Which draws the most current? (c) What happens to the brightness of A and B when C is unscrewed? (d) What happens to the brightness of B and C if A is unscrewed?

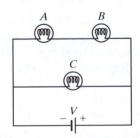

4. In the following circuit, you are to measure the potential difference across and the current through resistor R. (a) Indicate how you would connect the volt-

meter and ammeter in the circuit. (*b*) What resistances should an ideal voltmeter and an ideal ammeter have in order to make accurate measurements?

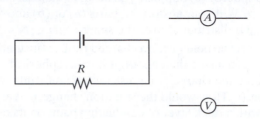

5.   (*a*) When resistors are connected in parallel, which of the following are the same for each: current, power, potential difference? (*b*) When resistors are connected in series, which of the following are the same for each: current, power, potential difference?

6.   Suppose bulb *A* in the circuit shown is unscrewed from its socket. (*a*) How do the brightnesses of the three remaining bulbs change? (*b*) How do these brightnesses compare with each other?

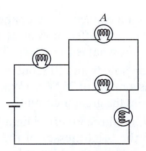

7.   Imagine that some node in a certain circuit does not obey Kirchhoff's first law—for instance, the currents into and out of the node differ by some positive amount. What happens to the potential at that node?

8.   When an *RC* circuit is connected to a battery, the circuit draws current until the capacitor is fully charged. Is the amount of energy stored in the capacitor larger than, equal to, or smaller than the energy delivered to the circuit by the battery? Explain.

MAGNETISM

1.   Can an electric or a magnetic field, *each constant in space and time*, be used to accomplish the actions described below? Explain your answers. Indicate if the answer is valid for any orientation of the field(s). Must any other condition be satisfied? (*a*) move a charged particle in a circle; (*b*) exert a force on a piece of dielectric; (*c*) increase the speed of a charged

particle; (*d*) accelerate a moving charged particle; (*e*) exert a force on an electron initially at rest.

2.   A compass is placed in a uniform magnetic field. What is the *net* force on the compass needle?

3.   Two very long, fixed wires cross each other perpendicularly. They do not touch but are close to each other, as shown. Equal currents flow in the wires, in the directions shown. Indicate the locus of points where the net magnetic field is zero.

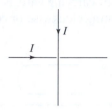

4.   A charged particle having a certain kinetic energy enters a static magnetic field. If no other forces act on the particle while it is within the field, it has the same kinetic energy on leaving the field. Why?

5.   A metallic bar moves to the right in a uniform magnetic field that points out of the page, as shown. (*a*) Is the electric field inside the bar zero? Explain. (*b*) Is an external force required to keep the bar moving with constant velocity? Explain.

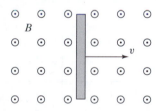

6.   In the following figure, the bar moves to the right with a velocity *v* in a uniform magnetic field that points out of the page. (*a*) Is the induced current clockwise or counterclockwise? (*b*) If the bar moves to the left, will the current reverse?

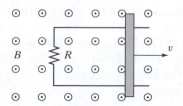

7. A very long line of positive charges moves in the direction of its long axis. (*a*) Is an electric field produced? (*b*) What about a magnetic field? (*c*) If these exist, sketch their direction(s). (*d*) Suppose we move a thin, neutral metal wire (consisting of static positive ions and free electrons) instead of the line of positive charges. Answer the same questions for this case.

8. In the following figure, a closed loop moves at a constant speed parallel to a long, straight, current-carrying wire. Is there a current in the loop? If so, is this current circulating clockwise or counterclockwise?

9. The square conducting sheet shown is moving at a uniform velocity v through a uniform magnetic field B that is perpendicular to the sheet. (*a*) What happens to free charges inside the sheet when it begins to move? What happens to these charges while the sheet is moving at a uniform velocity? (*b*) Do your answers to (*a*) depend on whether there are positive or negative (or both) free charges in the sheet? (*c*) Draw the equipotentials for the sheet in motion.

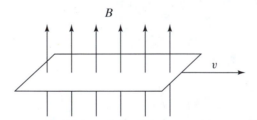

10. Charged particles passing through a bubble chamber leave tracks consisting of small hydrogen gas bubbles. These bubbles make visible the particles' trajectories. In the following figure, the magnetic field is directed into the page, and the tracks are in the plane of the page, in the directions indicated by the arrows. (*a*) Which of the tracks correspond to positively charged particles? (*b*) If all three particles have the same mass and charges of equal magnitude, which is moving the fastest? (*c*) If all three particles are moving with the same speed and have charges of equal magnitude, which has the greatest mass?

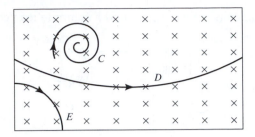

11. (*a*) Is it possible for a charged particle to move through a magnetic field without experiencing a force? (*b*) Is it possible for a charged particle to be at rest in the simultaneous presence of a magnetic and an electric field? (*c*) Is it possible for such a particle to move through the fields without experiencing a force? (*d*) What is the net force on a magnetic dipole in a uniform magnetic field?

INDUCTION AND MAXWELL'S EQUATIONS

1. In the following figure, a coil connected to a battery and a switch lies above a thin metallic ring. When the switch is opened, is the ring repelled by or attracted to the coil?

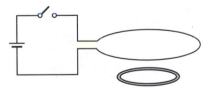

2. The following figure shows the two ways two identical inductors may be connected in series. In which case is the total inductance of the system larger (*i.e.*, in which case is the total "back-emf" of the series bigger)? Take into account the emf induced in each coil by the magnetic field of the other.

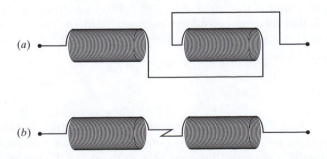

(*a*)

(*b*)

3. Two identical coils are wound around a common iron core, as in the following figure.

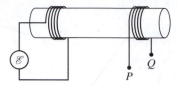

A sinusoidally varying emf of the form $\mathcal{E} = \mathcal{E}_0 \sin \omega t$ is applied to the left coil while the circuit on the right is left open (*i.e.*, the points P and Q are not connected). (*a*) Is the induced emf across the right coil (between the points P and Q) in phase, 90° ahead of, 90° behind, or 180° out of phase with the applied emf $\mathcal{E}$? Explain. (*b*) Does the induced emf across the right coil induce an emf in the left coil? What happens if a resistor is used to connect the points P and Q? Explain.

4. (*a*) A brass ring is placed on top of a coil of wire. If a switch to a source of direct current is closed and charges start to flow in the coil, the ring springs upward. Explain. (*b*) The ring is then placed atop the coil once again, and the switch opened. The current in the coil rapidly dies out. What happens to the ring now? Explain.

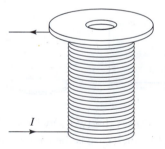

5. Write down the Maxwell equation most directly associated with the following statements and give a brief reason for each answer. Be sure to define the quantities used in each equation. (*a*) An alternating emf is induced in a coil that rotates in a uniform magnetic field. (*b*) The lines of the magnetic field circle round a steady current. (*c*) The static electric field inside a conductor is zero.

AC CIRCUITS

1. You are given a small box equipped with two electrical terminals. Upon applying an alternating current to the two terminals, you notice that the instantaneous potential difference across them and the instantaneous cur-

rent through the box are in phase. (*a*) What conclusions can you make about the resistance and/or the capacitive and inductive reactance of the box? (*b*) If you lower the frequency, will the potential difference and the current remain in phase?

2. In order to run an ordinary household appliance from the cigarette lighter of a car, you can buy a device that converts the car battery's 12-V dc to 110-V ac. One part of the circuit in such a device makes the voltage alternating; another uses a transformer to increase the voltage. Does it make a difference in which order these two operations are carried out?

3. The same ac voltage is applied across a resistor and a capacitor, and the same rms current is measured through each. Do both dissipate the same amount of energy? Explain.

4. In a series RLC circuit, the current lags the applied voltage. Is the peak voltage greater across the capacitor or across the inductor?

5. Fluorescent light fixtures use an inductor, called the ballast, to limit the current through the lamp. Why is an inductor used instead of a resistor?

6. A light bulb designed to operate at 120 V (rms) is connected in series with an inductor, a capacitor, and a 120-V (rms) ac source. (*a*) Is it possible for this bulb to be as bright as an identical one connected directly to the source? (*b*) Is it possible for this bulb to be brighter than an identical one connected directly to the source?

7. A circuit containing a coil, a resistor, and a battery has achieved a steady state; the current has reached a constant value. (*a*) Does the coil have an inductance? (*b*) Does the coil affect the value of the current in the circuit?

8. A lightbulb and an inductor are connected in parallel to an alternating source of emf. (*a*) How does the brightness of the bulb vary as the frequency of the ac source is increased from zero? (*b*) If the inductor is replaced by a capacitor, how does your answer change?

9. The sum of the peak voltages across the elements in a series RLC circuit is usually greater than the peak applied voltage. Why? Does this violate Kirchhoff's voltage law?

10. Is the ac voltage applied to a circuit always in phase with the current through a resistor in that circuit?

11. If a capacitor is connected to a sinusoidally alternating source of emf, does the displacement current between the capacitor plates lead or lag behind the emf?

12. The following figure shows a phasor diagram for a series RLC circuit. (*a*) Identify each of the three components 1, 2, and 3 (*e.g.*, $1 = R, 2 = L, 3 = C$). (*b*) Is the driving frequency above or below resonance?

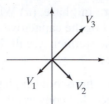

13. Step-up transformers are used to increase potential differences. The potential difference, however, represents energy per unit charge, and so a step-up transformer increases the energy per unit charge. How is this possible without violating the law of conservation of energy?

OPTICS

1. A thin converging lens is used to form a real image of a nearby object. If the object is moved closer to the lens, a new real image is observed. Does the new image differ from the old one (*a*) in position relative to the lens, and (*b*) in size? If it does, describe how.

2. By changing the separation distance between lens and film, a camera can focus on subjects at a variety of distances. Suppose the proper lens-film separation for taking an in-focus picture of a distant object such as the moon is d_o. To take an in-focus picture of a nearby object, will the proper lens-film separation be greater than, equal to, or less than d_o? Explain using diagrams.

3. (*a*) Does light traveling from one medium to another always bend toward the normal? (*b*) As light travels across the interface between two media, does its frequency change? (*c*) Its wavelength? (*d*) Its velocity? Explain your answers.

MODERN PHYSICS

1. Many houses have black asphalt roofs. From an energy-saving point of view, is this the best roof finish on (*a*) a hot summer day and (*b*) a cold winter night? In either case, if not, what type of roof would be better?

2. In order for an atom to emit light, must it first be ionized? Explain.

3. Do the emission and absorption spectra obtained from the same gas at room temperature contain the same number of lines? Explain.

4. Can a photon—*any* photon—be absorbed by a free electron? (*Hint:* consider conservation of momentum and energy.)

5. If an electron and a proton are accelerated from rest through the same potential difference, which has the longer wavelength?

6. An electron in a hydrogen atom is in its ground state ($n = 1$). (a) If radiation having a frequency greater than $(E_3 - E_1)/h$ but less than $(E_4 - E_1)/h$ is incident, what happens? (b) If instead a beam of electrons having kinetic energy greater than $(E_3 - E_1)$ but less than $(E_4 - E_1)$ is used, what happens?

7. Would a beam of protons in a "proton microscope" exhibit more or less diffraction than a beam of electrons of the same speed in an electron microscope? Explain.

8. One thousand photons uniformly distributed in wavelength between 1100 and 1600 nm exert pressure on a surface they strike. Suppose a prism is used to remove all the photons having wavelengths below 1350 nm. Will the radiation pressure be reduced by (a) less than half, (b) half, or (c) more than half? Explain.

APPENDIX:
CD-ROM INSTRUCTIONS

The CD-ROM at the back of *Peer Instruction: A User's Manual* contains all the resource material—Chapters 7–12 and the two questionnaires in Chapter 3—in Adobe PDF (Portable Document Format) files. These files can be viewed, navigated, and printed in their original form independent of computer platform and operating system.

Adobe Acrobat Reader is needed to use these files. This software is free and can be downloaded in just a few minutes from Adobe's web site. Point your web browser to http://www.adobe.com and follow the instructions for installing Adobe Acrobat Reader on your type of computer.

Some important notes about the PDF files on your CD-ROM

We have reformatted the material in the resource chapters to make it as easy as possible for you to use the resources in class. Using the CD-ROM, all the materials can be printed onto $8\frac{1}{2} \times 11$ format or transparencies.

Each of the *ConcepTests* and Reading Quizzes has been re-sized and re-formatted for use as an overhead or handout master.

You can also electronically copy the material (with the exception of the *Force Concept Inventory* and *Mechanics Baseline Test*), paste it into other applications, and re-organize or modify it to suit your needs.

License

You may use the materials for educational purposes as described above and you have permission to copy, or have copied, the resource materials in the PDF files for your class. No portion of the materials, whether in original or in altered form, may be otherwise distributed, transmitted in any form, or included in other documents without express written permission from the publisher or, in the case of the *Force Concept Inventory* and *Mechanics Baseline Test*, the copyright holders.

INDEX

Note: CEQ refers to Conceptual Exam Questions; CT refers to *ConcepTests*; and RQ refers to Reading Quizzes. The first number or set of numbers following an abbreviation refers to the question number(s) for that topic; the second number or set of numbers (after the colon) indicates the page(s) on which the question(s) are found. For example, RQ1: 23 refers to Reading Quiz 1 on page 23; CT1-4: 163-164 refers to *ConcepTest* questions 1-4 on pages 163-164.

Note: CEQ refers to Conceptual Exam Questions; CT refers to *ConcepTests*; and RQ refers to Reading Quizzes. The first number or set of numbers following an abbreviation refers to the question number(s) for that topic; the second number or set of numbers (after the colon) indicates the page(s) on which the question(s) are found. For example, RQ1: 23 refers to Reading Quiz 1 on page 23; CT1-4: 163-164 refers to *ConcepTest* questions 1-4 on pages 163-164.

Note: CEQ refers to Conceptual Exam Questions; CT refers to *ConcepTests*; and RQ refers to Reading Quizzes. The first number or set of numbers following an abbreviation refers to the question number(s) for that topic; the second number or set of numbers (after the colon) indicates the page(s) on which the question(s) are found. For example, RQ1: 23 refers to Reading Quiz 1 on page 23; CT1-4: 163-164 refers to *ConcepTest* questions 1-4 on pages 163-164.

Note: CEQ refers to Conceptual Exam Questions; CT refers to *ConcepTests*; and RQ refers to Reading Quizzes. The first number or set of numbers following an abbreviation refers to the question number(s) for that topic; the second number or set of numbers (after the colon) indicates the page(s) on which the question(s) are found. For example, RQ1: 23 refers to Reading Quiz 1 on page 23; CT1-4: 163-164 refers to *ConcepTest* questions 1-4 on pages 163-164.

MORE PRAISE FOR *PEER INSTRUCTION: A USER'S MANUAL*

"Reform must be bold yet practical, and Mazur's approach satisfies these two conditions. Those of us who care about student learning will embrace this approach."

Robert S. Weidman
Michigan Tech University

"After reading the *Peer Instruction* manual, I am more confident I can get this approach started successfully in my class."

William J. Sturrus
Youngstown State University

"This is the real thing. Students are starved for it."

Peter Heller
Brandeis University

"We have learned the value of peer instruction and active involvement of students in the classroom."

George W. Parker
North Carolina State University

"This manual is well-written, highly-focused, and useful. Mazur articulates a number of well-known problems associated with reaching today's students and gives a useful prescription for solution."

Arthur Z. Kovacs
Rochester Institute of Technology